Praise for *Essential Math for Data Science*

In the cacophony that is the current data science education landscape, this book stands out as a resource with many clear, practical examples of the fundamentals of what it takes to understand and build with data. By explaining the basics, this book allows the reader to navigate any data science work with a sturdy mental framework of its building blocks.

—*Vicki Boykis, Senior Machine Learning Engineer at Tumblr*

Data science is built on linear algebra, probability theory, and calculus. Thomas Nield expertly guides us through all of those topics—and more—to build a solid foundation for understanding the mathematics of data science.

—*Mike X Cohen, sincXpress*

As data scientists, we use sophisticated models and algorithms daily. This book swiftly demystifies the math behind them, so they are easier to grasp and implement.

—*Siddharth Yadav, freelance data scientist*

I wish I had access to this book earlier! Thomas Nield does such an amazing job breaking down complex math topics in a digestible and engaging way. A refreshing approach to both math and data science—seamlessly explaining fundamental math concepts and their immediate applications in machine learning. This book is a must-read for all aspiring data scientists.

—*Tatiana Ediger, freelance data scientist and course developer and instructor*

Essential Math for Data Science

Take Control of Your Data with Fundamental
Linear Algebra, Probability, and Statistics

Thomas Nield

Beijing · Boston · Farnham · Sebastopol · Tokyo

Essential Math for Data Science

by Thomas Nield

Published by O'Reilly Media, Inc., 1005 Gravenstein Highway North, Sebastopol, CA 95472.

O'Reilly books may be purchased for educational, business, or sales promotional use. Online editions are also available for most titles (*http://oreilly.com*). For more information, contact our corporate/institutional sales department: 800-998-9938 or *corporate@oreilly.com*.

Acquisitions Editor: Jessica Haberman
Development Editor: Jill Leonard
Production Editor: Kristen Brown
Copyeditor: Piper Editorial Consulting, LLC
Proofreader: Shannon Turlington

Indexer: Potomac Indexing, LLC
Interior Designer: David Futato
Cover Designer: Karen Montgomery
Illustrator: Kate Dullea

June 2022: First Edition

Revision History for the First Edition
2022-05-26: First Release
2024-06-13: Second Release

See *http://oreilly.com/catalog/errata.csp?isbn=9781098102937* for release details.

The O'Reilly logo is a registered trademark of O'Reilly Media, Inc. *Essential Math for Data Science*, the cover image, and related trade dress are trademarks of O'Reilly Media, Inc.

978-1-098-10293-7

[LSI]

Table of Contents

Preface

In the past 10 years or so, there has been a growing interest in applying math and statistics to our everyday work and lives. Why is that? Does it have to do with the accelerated interest in "data science," which Harvard Business Review called "the Sexiest Job of the 21st Century" (*https://oreil.ly/GslO6*)? Or is it the promise of machine learning and "artificial intelligence" changing our lives? Is it because news headlines are inundated with studies, polls, and research findings but unsure how to scrutinize such claims? Or is it the promise of "self-driving" cars and robots automating jobs in the near future?

I will make the argument that the disciplines of math and statistics have captured mainstream interest because of the growing availability of data, and we need math, statistics, and machine learning to make sense of it. Yes, we do have scientific tools, machine learning, and other automations that call to us like sirens. We blindly trust these "black boxes," devices, and softwares; we do not understand them but we use them anyway.

While it is easy to believe computers are smarter than we are (and this idea is frequently marketed), the reality cannot be more the opposite. This disconnect can be precarious on so many levels. Do you really want an algorithm or AI performing criminal sentencing or driving a vehicle, but nobody including the developer can explain why it came to a specific decision? Explainability is the next frontier of statistical computing and AI. This can begin only when we open up the black box and uncover the math.

You may also ask how can a developer not know how their own algorithm works? We will talk about that in the second half of the book when we discuss machine learning techniques and emphasize why we need to understand the math behind the black boxes we build.

To another point, the reason data is being collected on a massive scale is largely due to connected devices and their presence in our everyday lives. We no longer solely use the internet on a desktop or laptop computer. We now take it with us in our smartphones, cars, and household devices. This has subtly enabled a transition over the past two decades. Data has now evolved from an operational tool to something that is collected and analyzed for less-defined objectives. A smartwatch is constantly collecting data on our heart rate, breathing, walking distance, and other markers. Then it uploads that data to a cloud to be analyzed alongside other users. Our driving habits are being collected by computerized cars and being used by manufacturers to collect data and enable self-driving vehicles. Even "smart toothbrushes" are finding their way into drugstores, which track brushing habits and store that data in a cloud. Whether smart toothbrush data is useful and essential is another discussion!

All of this data collection is permeating every corner of our lives. It can be overwhelming, and a whole book can be written on privacy concerns and ethics. But this availability of data also creates opportunities to leverage math and statistics in new ways and create more exposure outside academic environments. We can learn more about the human experience, improve product design and application, and optimize commercial strategies. If you understand the ideas presented in this book, you will be able to unlock the value held in our data-hoarding infrastructure. This does not imply that data and statistical tools are a silver bullet to solve all the world's problems, but they have given us new tools that we can use. Sometimes it is just as valuable to recognize certain data projects as rabbit holes and realize efforts are better spent elsewhere.

This growing availability of data has made way for data science and machine learning to become in-demand professions. We define essential math as an exposure to probability, linear algebra, statistics, and machine learning. If you are seeking a career in data science, machine learning, or engineering, these topics are necessary. I will throw in just enough college math, calculus, and statistics necessary to better understand what goes in the black box libraries you will encounter.

With this book, I aim to expose readers to different mathematical, statistical, and machine learning areas that will be applicable to real-world problems. The first four chapters cover foundational math concepts including practical calculus, probability, linear algebra, and statistics. The last three chapters will segue into machine learning. The ultimate purpose of teaching machine learning is to integrate everything we learn and demonstrate practical insights in using machine learning and statistical libraries beyond a black box understanding.

The only tool needed to follow examples is a Windows/Mac/Linux computer and a Python 3 environment of your choice. The primary Python libraries we will need are `numpy`, `scipy`, `sympy`, and `sklearn`. If you are unfamiliar with Python, it is a friendly and easy-to-use programming language with massive learning resources behind it. Here are some I recommend:

Data Science from Scratch, 2nd Edition by Joel Grus (O'Reilly)

> The second chapter of this book has the best crash course in Python I have encountered. Even if you have never written code before, Joel does a fantastic job getting you up and running with Python effectively in the shortest time possible. It is also a great book to have on your shelf and to apply your mathematical knowledge!

Python for the Busy Java Developer by Deepak Sarda (Apress)

> If you are a software engineer coming from a statically-typed, object-oriented programming background, this is the book to grab. As someone who started programming with Java, I have a deep appreciation for how Deepak shares Python features and relates them to Java developers. If you have done .NET, C++, or other C-like languages you will probably learn Python effectively from this book as well.

This book will not make you an expert or give you PhD knowledge. I do my best to avoid mathematical expressions full of Greek symbols and instead strive to use plain English in its place. But what this book will do is make you more comfortable talking about math and statistics, giving you *essential* knowledge to navigate these areas successfully. I believe the widest path to success is not having deep, specialized knowledge in one topic, but instead having exposure and practical knowledge across several topics. That is the goal of this book, and you will learn just enough to be dangerous and ask those once-elusive critical questions.

So let's get started!

Conventions Used in This Book

The following typographical conventions are used in this book:

Italic

> Indicates new terms, URLs, email addresses, filenames, and file extensions.

`Constant width`

> Used for program listings, as well as within paragraphs to refer to program elements such as variable or function names, databases, data types, environment variables, statements, and keywords.

`Constant width bold`

> Shows commands or other text that should be typed literally by the user.

`Constant width italic`

> Shows text that should be replaced with user-supplied values or by values determined by context.

This element signifies a tip or suggestion.

This element signifies a general note.

This element indicates a warning or caution.

Using Code Examples

Supplemental material (code examples, exercises, etc.) is available for download at *https://github.com/thomasnield/machine-learning-demo-data*.

If you have a technical question or a problem using the code examples, please send email to *bookquestions@oreilly.com*.

This book is here to help you get your job done. In general, if example code is offered with this book, you may use it in your programs and documentation. You do not need to contact us for permission unless you're reproducing a significant portion of the code. For example, writing a program that uses several chunks of code from this book does not require permission. Selling or distributing examples from O'Reilly books does require permission. Answering a question by citing this book and quoting example code does not require permission. Incorporating a significant amount of example code from this book into your product's documentation does require permission.

We appreciate, but generally do not require, attribution. An attribution usually includes the title, author, publisher, and ISBN. For example: "*Essential Math for Data Science* by Thomas Nield (O'Reilly). Copyright 2022 Thomas Nield, 978-1-098-10293-7."

If you feel your use of code examples falls outside fair use or the permission given above, feel free to contact us at *permissions@oreilly.com*.

O'Reilly Online Learning

 For more than 40 years, *O'Reilly Media* has provided technology and business training, knowledge, and insight to help companies succeed.

Our unique network of experts and innovators share their knowledge and expertise through books, articles, and our online learning platform. O'Reilly's online learning platform gives you on-demand access to live training courses, in-depth learning paths, interactive coding environments, and a vast collection of text and video from O'Reilly and 200+ other publishers. For more information, visit *https://oreilly.com*.

How to Contact Us

Please address comments and questions concerning this book to the publisher:

> O'Reilly Media, Inc.
> 1005 Gravenstein Highway North
> Sebastopol, CA 95472
> 800-889-8969 (in the United States or Canada)
> 707-827-7019 (international or local)
> 707-829-0104 (fax)
> support@oreilly.com
> *https://oreilly.com/about/contact.html*

We have a web page for this book, where we list errata, examples, and any additional information. You can access this page at *https://oreil.ly/essentialMathDataSci*.

For news and information about our books and courses, visit *https://oreilly.com*.

Find us on LinkedIn: *https://linkedin.com/company/oreilly-media*

Watch us on YouTube: *https://youtube.com/oreillymedia*

Acknowledgments

This book was over a year's worth of efforts from many people. First, I want to thank my wife Kimberly for her support while I wrote this book, especially as we raised our son, Wyatt, to his first birthday. Kimberly is an amazing wife and mother, and everything I do now is for my son and our family's better future.

I want to thank my parents for teaching me to struggle past my limits and to never throw in the towel. Given this book's topic, I'm glad they encouraged me to take

calculus seriously in high school and college, and nobody can write a book without regularly exiting their comfort zone.

I want to thank the amazing team of editors and staff at O'Reilly who have continued to open doors since I wrote my first book on SQL in 2015. Jill and Jess have been amazing to work with in getting this book written and published, and I'm grateful that Jess thought of me when this topic came up.

I want to thank my colleagues at University of Southern California in the Aviation Safety and Security program. To have been given the opportunity to pioneer concepts in artificial intelligence system safety has taught me insights few people have, and I look forward to seeing what we continue to accomplish in the years to come. Arch, you continue to amaze me and I worry the world will stop functioning the day you retire.

Lastly, I want to thank my brother Dwight Nield and my friend Jon Ostrower, who are partners in my venture, Yawman Flight. Bootstrapping a startup is hard, and their help has allowed precious bandwidth to write this book. Jon brought me onboard at USC and his tireless accomplishments in the aviation journalism world are nothing short of remarkable (look him up!). It is an honor that they are as passionate as I am about an invention I started in my garage, and I don't think I could bring it to the world without them.

To anybody I have missed, thank you for the big and small things you have done. More often than not, I've been rewarded for being curious and asking questions. I do not take that for granted. As Ted Lasso said, "Be curious, not judgmental."

Basic Math and Calculus Review

We will kick off the first chapter covering what numbers are and how variables and functions work on a Cartesian system. We will then cover exponents and logarithms. After that, we will learn the two basic operations of calculus: derivatives and integrals.

Before we dive into the applied areas of essential math such as probability, linear algebra, statistics, and machine learning, we should probably review a few basic math and calculus concepts. Before you drop this book and run screaming, do not worry! I will present how to calculate derivatives and integrals for a function in a way you were probably not taught in college. We have Python on our side, not a pencil and paper. Even if you are not familiar with derivatives and integrals, you still do not need to worry.

I will make these topics as tight and practical as possible, focusing only on what will help us in later chapters and what falls under the "essential math" umbrella.

This Is Not a Full Math Crash Course!

This is by no means a comprehensive review of high school and college math. If you want that, a great book to check out is *No Bullshit Guide to Math and Physics* by Ivan Savov (pardon my French). The first few chapters contain the best crash course on high school and college math I have ever seen. The book *Mathematics 1001* by Dr. Richard Elwes has some great content as well, and in bite-sized explanations.

Number Theory

What are numbers? I promise to not be too philosophical in this book, but are numbers not a construct we have defined? Why do we have the digits 0 through 9, and not have more digits than that? Why do we have fractions and decimals and not just whole numbers? This area of math where we muse about numbers and why we designed them a certain way is known as number theory.

Number theory goes all the way back to ancient times, when mathematicians studied different number systems, and it explains why we have accepted them the way we do today. Here are different number systems that you may recognize:

Natural numbers

These are the numbers 1, 2, 3, 4, 5...and so on. Only positive numbers are included here, and they are the earliest known system. Natural numbers are so ancient cavemen scratched tally marks on bones and cave walls to keep records.

Whole numbers

Adding to natural numbers, the concept of "0" was later accepted; we call these "whole numbers." The Babylonians also developed the useful idea for place-holding notation for empty "columns" on numbers greater than 9, such as "10," "1,000," or "1,090." Those zeros indicate no value occupying that column.

Integers

Integers include positive and negative natural numbers as well as 0. We may take them for granted, but ancient mathematicians deeply distrusted the idea of negative numbers. But when you subtract 5 from 3, you get −2. This is useful especially when it comes to finances where we measure profits and losses. In 628 AD, an Indian mathematician named Brahmagupta showed why negative numbers were necessary for arithmetic to progress with the quadratic formula, and therefore integers became accepted.

Rational numbers

Any number that you can express as a fraction, such as 2/3, is a rational number. This includes all finite decimals and integers since they can be expressed as fractions, too, such as 687/100 = 6.87 and 2/1 = 2, respectively. They are called *rational* because they are *ratios*. Rational numbers were quickly deemed necessary because time, resources, and other quantities could not always be measured in discrete units. Milk does not always come in gallons. We may have to measure it as parts of a gallon. If I run for 12 minutes, I cannot be forced to measure in whole miles when in actuality I ran 9/10 of a mile.

Irrational numbers

Irrational numbers cannot be expressed as a fraction. This includes the famous π, square roots of certain numbers like $\sqrt{2}$, and Euler's number e, which we will learn about later. These numbers have an infinite number of decimal digits, such as 3.141592653589793238462…

There is an interesting history behind irrational numbers. The Greek mathematician Pythagoras believed all numbers are rational. He believed this so fervently, he made a religion that prayed to the number 10. "Bless us, divine number, thou who generated gods and men!" he and his followers would pray (why "10" was so special, I do not know). There is a legend that one of his followers, Hippasus, proved not all numbers are rational simply by demonstrating the square root of 2. This severely messed with Pythagoras's belief system, and he responded by drowning Hippasus at sea.

Regardless, we now know not all numbers are rational.

Real numbers

Real numbers include rational as well as irrational numbers. In practicality, when you are doing any data science work you can treat any decimals you work with as real numbers.

Complex and imaginary numbers

You encounter this number type when you take the square root of a negative number. While imaginary and complex numbers have relevance in certain types of problems, we will mostly steer clear of them.

In data science, you will find most (if not all) of your work will be using whole numbers, natural numbers, integers, and real numbers. Imaginary numbers may be encountered in more advanced use cases such as matrix decomposition, which we will touch on in Chapter 4.

Complex and Imaginary Numbers

If you do want to learn about imaginary numbers, there is a great playlist *Imaginary Numbers are Real* on YouTube (*https://oreil.ly/ bvyIq*).

Order of Operations

Hopefully, you are familiar with *order of operations*, which is the order you solve each part of a mathematical expression. As a brief refresher, recall that you evaluate components in parentheses, followed by exponents, then multiplication, division, addition, and subtraction. After that order, operations are then performed left-to-right. You can remember the order of operations by the mnemonic device PEMDAS (Please Excuse My Dear Aunt Sally), which corresponds to the ordering parentheses, exponents, multiplication, division, addition, and subtraction.

Take for example this expression:

$$2 \times \frac{(3 + 2)^2}{5} - 4$$

First we evaluate the parentheses (3 + 2), which equals 5:

$$2 \times \frac{(5)^2}{5} - 4$$

Next we solve the exponent, which we can see is squaring that 5 we just summed. That is 25:

$$2 \times \frac{25}{5} - 4$$

Next up we have multiplication and division. Let's go ahead and multiply the 2 with the $\frac{25}{5}$, yielding $\frac{50}{5}$:

$$\frac{50}{5} - 4$$

Next we will perform the division, dividing 50 by 5, which will yield 10:

$$10 - 4$$

And finally, we perform any addition and subtraction. Of course, $10 - 4$ is going to give us 6:

$$10 - 4 = 6$$

Sure enough, if we were to express this in Python we would print a value of 6.0 as shown in Example 1-1.

Example 1-1. Solving an expression in Python

```
my_value = 2 * (3 + 2)**2 / 5 - 4

print(my_value) # prints 6.0
```

This may be elementary but it is still critical. In code, even if you get the correct result without them, it is a good practice to liberally use parentheses in complex expressions so you establish control of the evaluation order.

Here I group the fractional part of my expression in parentheses, helping to set it apart from the rest of the expression in Example 1-2.

Example 1-2. Making use of parentheses for clarity in Python

```
my_value = 2 * ((3 + 2)**2 / 5) - 4

print(my_value) # prints 6.0
```

While both examples are technically correct, the latter is more clear to us easily confused humans. If you or someone else makes changes to your code, the parentheses provide an easy reference of operation order as you make changes. This provides a line of defense against code changes to prevent bugs as well.

Variables

If you have done some scripting with Python or another programming language, you have an idea what a variable is. In mathematics, a *variable* is a named placeholder for an unspecified or unknown number.

You may have a variable x representing any real number, and you can multiply that variable without declaring what it is. In Example 1-3 we take a variable input x from a user and multiply it by 3.

Example 1-3. A variable in Python that is then multiplied

```
x = int(input("Please input a number\n"))

product = 3 * x

print(product)
```

There are some standard variable names for certain variable types. If these variable names and concepts are unfamiliar, no worries! But some readers might recognize we use theta θ to denote angles and beta β for a parameter in a linear regression. Greek symbols make awkward variable names in Python, so we would likely name these variables theta and beta in Python as shown in Example 1-4.

Example 1-4. Greek variable names in Python

```
beta = 1.75
theta = 30.0
```

Note also that variable names can be subscripted so that several instances of a variable name can be used. For practical purposes, just treat these as separate variables. If you encounter variables x_1, x_2, and x_3, just treat them as three separate variables as shown in Example 1-5.

Example 1-5. Expressing subscripted variables in Python

```
x1 = 3  # or x_1 = 3
x2 = 10 # or x_2 = 10
x3 = 44 # or x_3 = 44
```

Functions

Functions are expressions that define relationships between two or more variables. More specifically, a function takes *input variables* (also called *domain variables* or *independent variables*), plugs them into an expression, and then results in an *output variable* (also called *dependent variable*).

Take this simple linear function:

$$y = 2x + 1$$

For any given x-value, we solve the expression with that x to find y. When $x = 1$, then $y = 3$. When $x = 2$, $y = 5$. When $x = 3$, $y = 7$ and so on, as shown in Table 1-1.

Table 1-1. Different values for y = 2x + 1

x	2x + 1	y
0	2(0) + 1	1
1	2(1) + 1	3
2	2(2) + 1	5
3	2(3) + 1	7

Functions are useful because they model a predictable relationship between variables, such as how many fires y can we expect at x temperature. We will use linear functions to perform linear regressions in Chapter 5.

Another convention you may see for the dependent variable y is to explicitly label it a function of x, such as $f(x)$. So rather than express a function as $y = 2x + 1$, we can also express it as:

$$f(x) = 2x + 1$$

Example 1-6 shows how we can declare a mathematical function and iterate it in Python.

Example 1-6. Declaring a linear function in Python

```
def f(x):
    return 2 * x + 1


x_values = [0, 1, 2, 3]

for x in x_values:
    y = f(x)
    print(y)
```

When dealing with real numbers, a subtle but important feature of functions is they often have an infinite number of x-values and resulting y-values. Ask yourself this: how many x-values can we put through the function $y = 2x + 1$? Rather than just 0, 1, 2, 3…why not 0, 0.5, 1, 1.5, 2, 2.5, 3 as shown in Table 1-2?

Table 1-2. Different values for y = 2x + 1

x	2x + 1	y
0.0	2(0) + 1	1
0.5	2(.5) + 1	2
1.0	2(1) + 1	3
1.5	2(1.5) + 1	4
2.0	2(2) + 1	5
2.5	2(2.5) + 1	6
3.0	2(3) + 1	7

Or why not do quarter steps for x? Or 1/10 of a step? We can make these steps infinitely small, effectively showing $y = 2x + 1$ is a *continuous function*, where for every possible value of x there is a value for y. This segues us nicely to visualize our function as a line as shown in Figure 1-1.

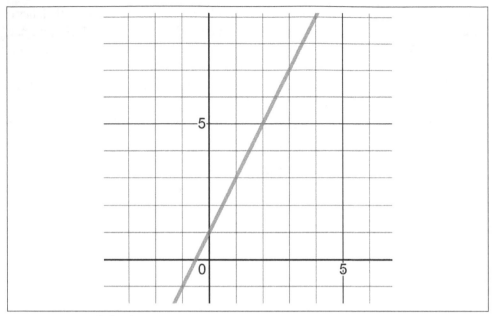

Figure 1-1. Graph for function y = 2x + 1

When we plot on a two-dimensional plane with two number lines (one for each variable) it is known as a *Cartesian plane*, *x-y plane*, or *coordinate plane*. We trace a given x-value and then look up the corresponding y-value, and plot the intersections as a line. Notice that due to the nature of real numbers (or decimals, if you prefer), there are an infinite number of x values. This is why when we plot the function $f(x)$ we get a continuous line with no breaks in it. There are an infinite number of points on that line, or any part of that line.

If you want to plot this using Python, there are a number of charting libraries from Plotly to matplotlib. Throughout this book we will use SymPy to do many tasks, and the first we will use is plotting a function. SymPy uses matplotlib so make sure you have that package installed. Otherwise it will print an ugly text-based graph to your console. After that, just declare the x variable to SymPy using `symbols()`, declare your function, and then plot it as shown in Example 1-7 and Figure 1-2.

Example 1-7. Charting a linear function in Python using SymPy

```
from sympy import *

x = symbols('x')
f = 2*x + 1
plot(f)
```

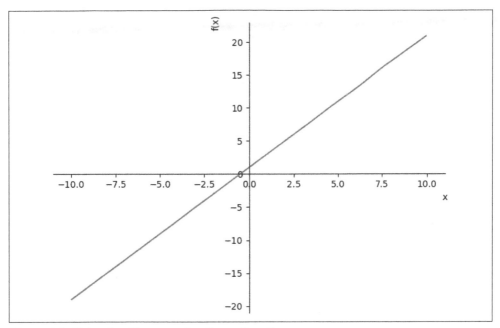

Figure 1-2. Using SymPy to graph a linear function

Example 1-8 and Figure 1-3 are another example showing the function $f(x) = x^2 + 1$.

Example 1-8. Charting a function involving an exponent

```
from sympy import *

x = symbols('x')
f = x**2 + 1
plot(f)
```

Note in Figure 1-3 we do not get a straight line but rather a smooth, symmetrical curve known as a parabola. It is continuous but not linear, as it does not produce values in a straight line. Curvy functions like this are mathematically harder to work with, but we will learn some tricks to make it not so bad.

Curvilinear Functions

When a function is continuous but curvy, rather than linear and straight, we call it a *curvilinear function*.

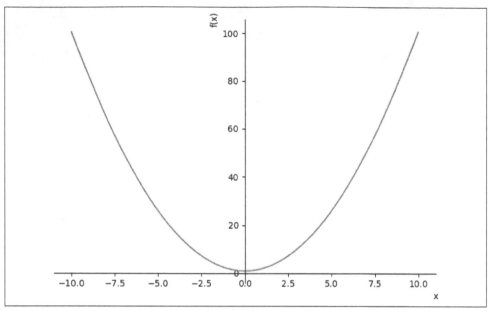

Figure 1-3. Using SymPy to graph a function involving an exponent

Note that functions utilize multiple input variables, not just one. For example, we can have a function with independent variables x and y. Note that y is not dependent like in previous examples.

$$f(x, y) = 2x + 3y$$

Since we have two independent variables (x and y) and one dependent variable (the output of $f(x,y)$), we need to plot this graph on three dimensions to produce a plane of values rather than a line, as shown in Example 1-9 and Figure 1-4.

Example 1-9. Declaring a function with two independent variables in Python

```
from sympy import *
from sympy.plotting import plot3d

x, y = symbols('x y')
f = 2*x + 3*y
plot3d(f)
```

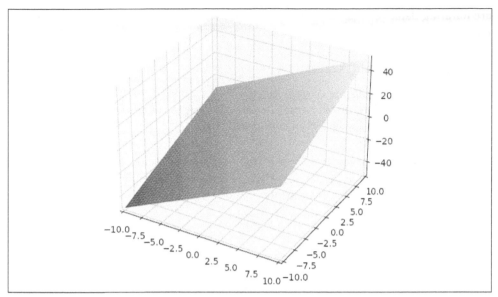

Figure 1-4. Using SymPy to graph a three-dimensional function

No matter how many independent variables you have, your function will typically output only one dependent variable. When you solve for multiple dependent variables, you will likely be using separate functions for each one.

Summations

I promised not to use equations full of Greek symbols in this book. However, there is one that is so common and useful that I would be remiss to not cover it. A *summation* is expressed as a sigma Σ and adds elements together.

For example, if I want to iterate the numbers 1 through 5, multiply each by 2, and sum them, here is how I would express that using a summation. Example 1-10 shows how to execute this in Python.

$$\sum_{i=1}^{5} 2i = (2)1 + (2)2 + (2)3 + (2)4 + (2)5 = 30$$

Example 1-10. Performing a summation in Python

```
summation = sum(2*i for i in range(1,6))
print(summation)
```

Note that i is a placeholder variable representing each consecutive index value we are iterating in the loop, which we multiply by 2 and then sum all together. When you are iterating data, you may see variables like x_i indicating an element in a collection at index i.

The range() function

Recall that the `range()` function in Python is end exclusive, meaning if you invoke `range(1,4)` it will iterate the numbers 1, 2, and 3. It excludes the 4 as an upper boundary.

It is also common to see n represent the number of items in a collection, like the number of records in a dataset. Here is one such example where we iterate a collection of numbers of size n, multiply each one by 10, and sum them:

$$\sum_{i=1}^{n} 10x_i$$

In Example 1-11 we use Python to execute this expression on a collection of four numbers. Note that in Python (and most programming languages in general) we typically reference items starting at index 0, while in math we start at index 1. Therefore, we shift accordingly in our iteration by starting at 0 in our `range()`.

Example 1-11. Summation of elements in Python

```
x = [1, 4, 6, 2]
n = len(x)

summation = sum(10*x[i] for i in range(0,n))
print(summation)
```

That is the gist of summation. In a nutshell, a summation Σ says, "add a bunch of things together," and uses an index i and a maximum value n to express each iteration feeding into the sum. We will see these throughout this book.

Summations in SymPy

Feel free to come back to this sidebar later when you learn more about SymPy. SymPy, which we use to graph functions, is actually a symbolic math library; we will talk about what this means later in this chapter. But note for future reference that a summation operation in SymPy is performed using the `Sum()` operator. In the following code, we iterate i from 1 through n, multiply each i, and sum them. But then we use the `subs()` function to specify n as 5, which will then iterate and sum all i elements from 1 through n:

```
from sympy import *

i,n = symbols('i n')

# iterate each element i from 1 to n,
# then multiply and sum
summation = Sum(2*i,(i,1,n))

# specify n as 5,
# iterating the numbers 1 through 5
up_to_5 = summation.subs(n, 5)
print(up_to_5.doit()) # 30
```

Note that summations in SymPy are lazy, meaning they do not automatically calculate or get simplified. So use the `doit()` function to execute the expression.

Exponents

Exponents multiply a number by itself a specified number of times. When you raise 2 to the third power (expressed as 2^3 using 3 as a superscript), that is multiplying three 2s together:

$$2^3 = 2 * 2 * 2 = 8$$

The *base* is the variable or value we are exponentiating, and the *exponent* is the number of times we multiply the base value. For the expression 2^3, 2 is the base and 3 is the exponent.

Exponents have a few interesting properties. Say we multiplied x^2 and x^3 together. Observe what happens next when I expand the exponents with simple multiplication and then consolidate into a single exponent:

$$x^2 x^3 = (x * x) * (x * x * x) = x^{2+3} = x^5$$

When we multiply exponents together with the same base, we simply add the exponents, which is known as the *product rule*. Let me emphasize that the base of all multiplied exponents must be the same for the product rule to apply.

Let's explore division next. What happens when we divide x^2 by x^5?

$$\frac{x^2}{x^5}$$

$$\frac{x * x}{x * x * x * x * x}$$

$$\frac{1}{x * x * x}$$

$$\frac{1}{x^3} = x^{-3}$$

As you can see, when we divide x^2 by x^5 we can cancel out two x's in the numerator and denominator, leaving us with $\frac{1}{x^3}$. When a factor exists in both the numerator and denominator, we can cancel out that factor.

What about the x^{-3}, you wonder? This is a good point to introduce negative exponents, which is another way of expressing an exponent operation in the denominator of a fraction. To demonstrate, $\frac{1}{x^3}$ is the same as x^{-3}:

$$\frac{1}{x^3} = x^{-3}$$

Tying back the product rule, we can see it applies to negative exponents, too. To get intuition behind this, let's approach this problem a different way. We can express this division of two exponents by making the "5" exponent of x^5 negative, and then multiplying it with x^2. When you add a negative number, it is effectively performing subtraction. Therefore, the exponent product rule summing the multiplied exponents still holds up as shown next:

$$\frac{x^2}{x^5} = x^2 \frac{1}{x^5} = x^2 x^{-5} = x^{2 + -5} = x^{-3}$$

Last but not least, can you figure out why any base with an exponent of 0 is 1?

$$x^0 = 1$$

The best way to get this intuition is to reason that any number divided by itself is 1. If you have $\frac{x^3}{x^3}$ it is algebraically obvious that reduces to 1. But that expression also evaluates to x^0:

$$1 = \frac{x^3}{x^3} = x^3 x^{-3} = x^{3 + -3} = x^0$$

By the transitive property, which states that if $a = b$ and $b = c$, then $a = c$, we know that $x^0 = 1$.

Simplify Expressions with SymPy

If you get uncomfortable with simplifying algebraic expressions, you can use the SymPy library to do the work for you. Here is how to simplify our previous example:

```
from sympy import *

x = symbols('x')
expr = x**2 / x**5
print(expr) # x**(-3)
```

Now what about fractional exponents? They are an alternative way to represent roots, such as the square root. As a brief refresher, a $\sqrt{4}$ asks "What number multiplied by itself will give me 4?" which of course is 2. Note here that $4^{1/2}$ is the same as $\sqrt{4}$:

$$4^{1/2} = \sqrt{4} = 2$$

Cubed roots are similar to square roots, but they seek a number multiplied by itself three times to give a result. A cubed root of 8 is expressed as $\sqrt[3]{8}$ and asks "What number multiplied by itself three times gives me 8?" This number would be 2 because $2 * 2 * 2 = 8$. In exponents a cubed root is expressed as a fractional exponent, and $\sqrt[3]{8}$ can be reexpressed as $8^{1/3}$:

$$8^{1/3} = \sqrt[3]{8} = 2$$

To bring it back full circle, what happens when you multiply the cubed root of 8 three times? This will undo the cubed root and yield 8. Alternatively, if we express the cubed root as fractional exponents $8^{1/3}$, it becomes clear we add the exponents together to get an exponent of 1. That also undoes the cubed root:

$$\sqrt[3]{8} * \sqrt[3]{8} * \sqrt[3]{8} = 8^{\frac{1}{3}} \times 8^{\frac{1}{3}} \times 8^{\frac{1}{3}} = 8^{\frac{1}{3} + \frac{1}{3} + \frac{1}{3}} = 8^1 = 8$$

And one last property: an exponent of an exponent will multiply the exponents together. This is known as the *power rule*. So $\left(8^3\right)^2$ would simplify to 8^6:

$$\left(8^3\right)^2 = 8^{3 \times 2} = 8^6$$

If you are skeptical why this is, try expanding it and you will see the sum rule makes it clear:

$$\left(8^3\right)^2 = 8^3 8^3 = 8^{3+3} = 8^6$$

Lastly, what does it mean when we have a fractional exponent with a numerator other than 1, such as $8^{\frac{2}{3}}$? Well, that is taking the cube root of 8 and then squaring it. Take a look:

$$8^{\frac{2}{3}} = \left(8^{\frac{1}{3}}\right)^2 = 2^2 = 4$$

And yes, irrational numbers can serve as exponents like 8^{π}, which is 687.2913. This may feel unintuitive, and understandably so! In the interest of time, we will not dive deep into this as it requires some calculus. But essentially, we can calculate irrational exponents by approximating with a rational number. This is effectively what computers do since they can compute to only so many decimal places anyway.

For example π has an infinite number of decimal places. But if we take the first 11 digits, 3.1415926535, we can approximate π as a rational number 31415926535 / 10000000000. Sure enough, this gives us approximately 687.2913, which should approximately match any calculator:

$$8^{\pi} \approx 8^{\frac{31415926535}{10000000000}} \approx 687.2913$$

Logarithms

A *logarithm* is a math function that finds a power for a specific number and base. It may not sound interesting at first, but it actually has many applications. From measuring earthquakes to managing volume on your stereo, the logarithm is found everywhere. It also finds its way into machine learning and data science a lot. As a matter of fact, logarithms will be a key part of logistic regressions in Chapter 6.

Start your thinking by asking "2 raised to *what power* gives me 8?" One way to express this mathematically is to use an x for the exponent:

$$2^x = 8$$

We intuitively know the answer, $x = 3$, but we need a more elegant way to express this common math operation. This is what the $log()$ function is for.

$$log_2 8 = x$$

As you can see in the preceding logarithm expression, we have a base 2 and are finding a power to give us 8. More generally, we can reexpress a variable exponent as a logarithm:

$$a^x = b$$
$$log_a b = x$$

Algebraically speaking, this is a way of isolating the x, which is important to solve for x. Example 1-12 shows how we calculate this logarithm in Python.

Example 1-12. Using the log function in Python

```
from math import log

# 2 raised to what power gives me 8?
x = log(8, 2)

print(x) # prints 3.0
```

When you do not supply a base argument to a `log()` function on a platform like Python, it will typically have a default base. In some fields, like earthquake measurements, the default base for the log is 10. But in data science the default base for the log is Euler's number e. Python uses the latter, and we will talk about e shortly.

Just like exponents, logarithms have several properties when it comes to multiplication, division, exponentiation, and so on. In the interest of time and focus, I will just present this in Table 1-3. The key idea to focus on is a logarithm finds an exponent for a given base to result in a certain number.

If you need to dive into logarithmic properties, Table 1-3 displays exponent and logarithm behaviors side-by-side that you can use for reference.

Table 1-3. Properties for exponents and logarithms

Operator	Exponent property	Logarithm property
Multiplication	$x^m \times x^n = x^{m+n}$	$log(a \times b) = log(a) + log(b)$
Division	$\frac{x^m}{x^n} = x^{m-n}$	$log\left(\frac{a}{b}\right) = log(a) - log(b)$
Exponentiation	$(x^m)^n = x^{mn}$	$log(a^n) = n \times log(a)$
Zero Exponent	$x^0 = 1$	$log(1) = 0$
Inverse	$x^{-1} = \frac{1}{x}$	$log(x^{-1}) = log\left(\frac{1}{x}\right) = -log(x)$

Euler's Number and Natural Logarithms

There is a special number that shows up quite a bit in math called Euler's number e. It is a special number much like Pi π and is approximately 2.71828. e is used a lot because it mathematically simplifies a lot of problems. We will cover e in the context of exponents and logarithms.

Euler's Number

Back in high school, my calculus teacher demonstrated Euler's number in several exponential problems. Finally I asked, "Mr. Nowe, what is e anyway? Where does it come from?" I remember never being fully satisfied with the explanations involving rabbit populations and other natural phenomena. I hope to give a more satisfying explanation here.

Why Euler's Number Is Used So Much

A property of Euler's number is its exponential function is a derivative to itself, which is convenient for exponential and logarithmic functions. We will learn about derivatives later in this chapter. In many applications where the base does not really matter, we pick the one that results in the simplest derivative, and that is Euler's number. That is also why it is the default base in many data science functions.

Here is how I like to discover Euler's number. Let's say you loan \$100 to somebody with 20% interest annually. Typically, interest will be compounded monthly, so the interest each month would be $.20/12 = .01666$. How much will the loan balance be after two years? To keep it simple, let's assume the loan does not require payments (and no payments are made) until the end of those two years.

Putting together the exponent concepts we learned so far (or perhaps pulling out a finance textbook), we can come up with a formula to calculate interest. It consists of a balance A for a starting investment P, interest rate r, time span t (number of years), and periods n (number of months in each year). Here is the formula:

$$A = P \times \left(1 + \frac{r}{n}\right)^{nt}$$

So if we were to compound interest every month, the loan would grow to \$148.69 as calculated here:

$$A = P \times \left(1 + \frac{r}{n}\right)^{nt}$$

$$100 \times \left(1 + \frac{.20}{12}\right)^{12 \times 2} = 148.6914618$$

If you want to do this in Python, try it out with the code in Example 1-13.

Example 1-13. Calculating compound interest in Python

```
from math import exp

p = 100
r = .20
t = 2.0
n = 12

a = p * (1 + (r/n))**(n * t)

print(a) # prints 148.69146179463576
```

But what if we compounded interest daily? What happens then? Change n to 365:

$$A = P \times \left(1 + \frac{r}{n}\right)^{nt}$$

$$100 \times \left(1 + \frac{.20}{365}\right)^{365 \times 2} = 149.1661279$$

Huh! If we compound our interest daily instead of monthly, we would earn 47.4666 cents more at the end of two years. If we got greedy why not compound every hour as shown next? Will that give us even more? There are 8,760 hours in a year, so set n to that value:

$$A = P \times \left(1 + \frac{r}{n}\right)^{nt}$$

$$100 \times \left(1 + \frac{.20}{8760}\right)^{8760 \times 2} = 149.1817886$$

Ah, we squeezed out roughly 2 cents more in interest! But are we experiencing a diminishing return? Let's try to compound every minute! Note that there are 525,600 minutes in a year, so let's set that value to n:

$$A = P \times \left(1 + \frac{r}{n}\right)^{nt}$$

$$100 \times \left(1 + \frac{.20}{525600}\right)^{525600 \times 2} = 149.1824584$$

OK, we are only gaining smaller and smaller fractions of a cent the more frequently we compound. So if I keep making these periods infinitely smaller to the point of compounding continuously, where does this lead?

Let me introduce you to Euler's number e, which is approximately 2.71828. Here is the formula to compound "continuously," meaning we are compounding nonstop:

$$A = P \times e^{rt}$$

Returning to our example, let's calculate the balance of our loan after two years if we compounded continuously:

$$A = P \times e^{rt}$$
$$A = 100 \times e^{.20 \times 2} = 149.1824698$$

This is not too surprising considering compounding every minute got us a balance of 149.1824584. That got us really close to our value of 149.1824698 when compounding continuously.

Typically you use e as an exponent base in Python, Excel, and other platforms using the exp() function. You will find that e is so commonly used, it is the default base for both exponent and logarithm functions.

Example 1-14 calculates continuous interest in Python using the exp() function.

Example 1-14. Calculating continuous interest in Python

```
from math import exp

p = 100 # principal, starting amount
r = .20 # interest rate, by year
t = 2.0 # time, number of years

a = p * exp(r*t)

print(a) # prints 149.18246976412703
```

So where do we derive this constant e? Compare the compounding interest formula and the continuous interest formula. They structurally look similar but have some differences:

$$A = P \times \left(1 + \frac{r}{n}\right)^{nt}$$
$$A = P \times e^{rt}$$

More technically speaking, e is the resulting value of the expression $\left(1 + \frac{1}{n}\right)^n$ as n forever gets bigger and bigger, thus approaching infinity. Try experimenting with increasingly large values for n. By making it larger and larger you will notice something:

$$\left(1 + \frac{1}{n}\right)^n$$

$$\left(1 + \frac{1}{100}\right)^{100} = 2.70481382942$$

$$\left(1 + \frac{1}{1000}\right)^{1000} = 2.71692393224$$

$$\left(1 + \frac{1}{10000}\right)^{10000} = 2.71814592682$$

$$\left(1 + \frac{1}{10000000}\right)^{10000000} = 2.71828169413$$

As you make n larger, there is a diminishing return and it converges approximately on the value 2.71828, which is our value e. You will find this e used not just in studying populations and their growth. It plays a key role in many areas of mathematics.

Later in the book, we will use Euler's number to build normal distributions in Chapter 3 and logistic regressions in Chapter 6.

Natural Logarithms

When we use e as our base for a logarithm, we call it a *natural logarithm*. Depending on the platform, we may use ln() instead of log() to specify a natural logarithm. So rather than express a natural logarithm expressed as $log_e 10$ to find the power raised on e to get 10, we would shorthand it as $ln(10)$:

$$log_e 10 = ln(10)$$

However, in Python, a natural logarithm is specified by the log() function. As discussed earlier, the default base for the log() function is e. Just leave the second argument for the base empty and it will default to using e as the base shown in Example 1-15.

Example 1-15. Calculating the natural logarithm of 10 in Python

```
from math import log

# e raised to what power gives us 10?
x = log(10)
```

```
print(x) # prints 2.302585092994046
```

We will use e in a number of places throughout this book. Feel free to experiment with exponents and logarithms using Excel, Python, Desmos.com, or any other calculation platform of your choice. Make graphs and get comfortable with what these functions look like.

Limits

As we have seen with Euler's number, some interesting ideas emerge when we forever increase or decrease an input variable and the output variable keeps approaching a value but never reaching it. Let's formally explore this idea.

Take this function, which is plotted in Figure 1-5:

$$f(x) = \frac{1}{x}$$

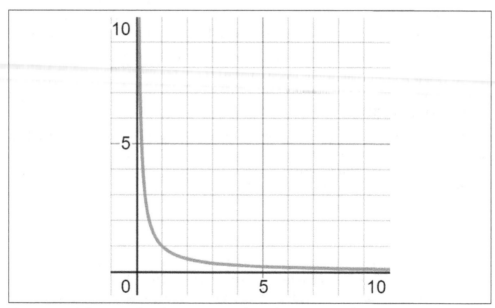

Figure 1-5. A function that forever approaches 0 but never reaches 0

We are looking only at positive x values. Notice that as x forever increases, $f(x)$ gets closer to 0. Fascinatingly, $f(x)$ never actually reaches 0. It just forever keeps getting closer.

Therefore the fate of this function is, as x forever extends into infinity, it will keep getting closer to 0 but never reach 0. The way we express a value that is forever being approached, but never reached, is through a limit:

$$\lim_{x \to \infty} \frac{1}{x} = 0$$

The way we read this is "as x approaches infinity, the function 1/x approaches 0 (but never reaches 0)." You will see this kind of "approach but never touch" behavior a lot, especially when we dive into derivatives and integrals.

Using SymPy, we can calculate what value we approach for $f(x) = \frac{1}{x}$ as x approaches infinity ∞ (Example 1-16). Note that ∞ is cleverly expressed in SymPy with oo.

Example 1-16. Using SymPy to calculate limits

```
from sympy import *

x = symbols('x')
f = 1 / x
result = limit(f, x, oo)

print(result) # 0
```

As you have seen, we discovered Euler's number e this way too. It is the result of forever extending n into infinity for this function:

$$\lim_{n \to \infty} \left(1 + \frac{1}{n}\right)^n = e = 2.71828169413...$$

Funnily enough, when we calculate Euler's number with limits in SymPy (shown in the following code), SymPy immediately recognizes it as Euler's number. We can call evalf() so we can actually display it as a number:

```
from sympy import *

n = symbols('n')
f = (1 + (1/n))**n
result = limit(f, n, oo)

print(result) # E
print(result.evalf()) # 2.71828182845905
```

Derivatives

Let's go back to talking about functions and look at them from a calculus perspective, starting with derivatives. A *derivative* tells the slope of a function, and it is useful to measure the rate of change at any point in a function.

Why do we care about derivatives? They are often used in machine learning and other mathematical algorithms, especially with gradient descent. When the slope is 0, that means we are at the minimum or maximum of an output variable. This concept will be useful later when we do linear regression (Chapter 5), logistic regression (Chapter 6), and neural networks (Chapter 7).

Let's start with a simple example. Let's take a look at the function $f(x) = x^2$ in Figure 1-6. How "steep" is the curve at *x = 2*?

Notice that we can measure "steepness" at any point in the curve, and we can visualize this with a tangent line. Think of a *tangent line* as a straight line that "just touches" the curve at a given point. It also provides the slope at a given point. You can crudely estimate a tangent line at a given x-value by creating a line intersecting that x-value and a *really close* neighboring x-value on the function.

Take *x = 2* and a nearby value *x = 2.1*, which when passed to the function $f(x) = x^2$ will yield *f*(2) = 4 and *f*(2.1) = 4.41 as shown in Figure 1-7. The resulting line that passes through these two points has a slope of 4.1.

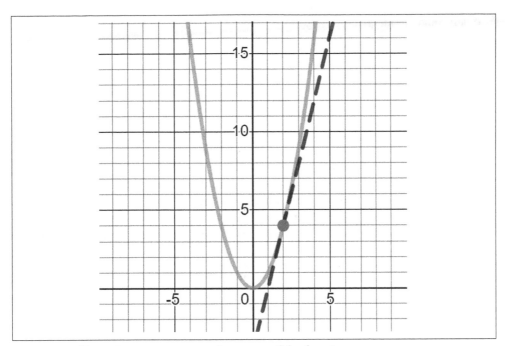

Figure 1-6. Observing steepness at a given part of the function

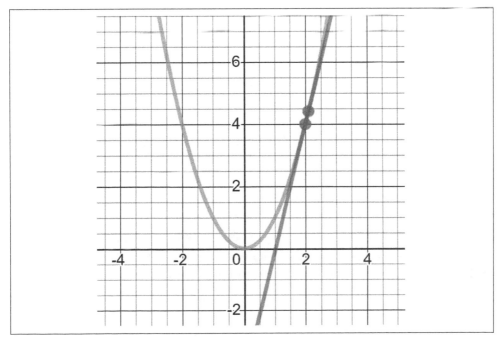

Figure 1-7. A crude way of calculating slope

You can quickly calculate the slope m between two points using the simple rise-over-run formula:

$$m = \frac{y_2 - y_1}{x_2 - x_1}$$

$$m = \frac{4.41 - 4.0}{2.1 - 2.0}$$

$$m = 4.1$$

If I made the x step between the two points even smaller, like $x = 2$ and $x = 2.00001$, which would result in $f(2) = 4$ and $f(2.00001) = 4.00004$, that would get *really* close to the actual slope of 4. So the smaller the step is to the neighboring value, the closer we get to the slope value at a given point in the curve. Like so many important concepts in math, we find something meaningful as we approach infinitely large or infinitely small values.

Example 1-17 shows a derivative calculator implemented in Python.

Example 1-17. A derivative calculator in Python

```
def derivative_x(f, x, step_size):
    m = (f(x + step_size) - f(x)) / ((x + step_size) - x)
    return m


def my_function(x):
    return x**2

slope_at_2 = derivative_x(my_function, 2, .00001)

print(slope_at_2) # prints 4.000010000000827
```

Now the good news is there is a cleaner way to calculate the slope anywhere on a function. We have already been using SymPy to plot graphs, but I will show you how it can also do tasks like derivatives using the magic of symbolic computation.

When you encounter an exponential function like $f(x) = x^2$ the derivative function will make the exponent a multiplier and then decrement the exponent by 1, leaving us with the derivative $\frac{d}{dx}x^2 = 2x$. The $\frac{d}{dx}$ indicates a *derivative with respect to x*, which says we are building a derivative targeting the x-value to get its slope. So if we want to find the slope at $x = 2$, and we have the derivative function, we just plug in that x-value to get the slope:

$$f(x) = x^2$$

$$\frac{d}{dx}f(x) = \frac{d}{dx}x^2 = 2x$$

$$\frac{d}{dx}f(2) = 2(2) = 4$$

If you intend to learn these rules to hand-calculate derivatives, there are plenty of calculus books for that. But there are some nice tools to calculate derivatives symbolically for you. The Python library SymPy is free and open source, and it nicely adapts to using the Python syntax. Example 1-18 shows how to calculate the derivative for $f(x) = x^2$ on SymPy.

Example 1-18. Calculating a derivative in SymPy

```
from sympy import *

# Declare 'x' to SymPy
x = symbols('x')

# Now just use Python syntax to declare function
f = x**2

# Calculate the derivative of the function
dx_f = diff(f)
print(dx_f) # prints 2*x
```

Wow! So by declaring variables using the `symbols()` function in SymPy, I can then proceed to use normal Python syntax to declare my function. After that I can use `diff()` to calculate the derivative function. In Example 1-19 we can then take our derivative function back to plain Python and simply declare it as another function.

Example 1-19. A derivative calculator in Python

```
def f(x):
    return x**2

def dx_f(x):
    return 2*x

slope_at_2 = dx_f(2.0)

print(slope_at_2) # prints 4.0
```

If you want to keep using SymPy, you can call the `subs()` function to swap the x variable with the value 2 as shown in Example 1-20.

Example 1-20. Using the substitution feature in SymPy

```
# Calculate the slope at x = 2
print(dx_f.subs(x,2)) # prints 4
```

Partial Derivatives

Another concept we will encounter in this book is *partial derivatives*, which we will use in Chapters 5, 6, and 7. These are derivatives of functions that have multiple input variables.

Think of it this way. Rather than finding the slope on a one-dimensional function, we have slopes with respect to multiple variables in several directions. For each given variable derivative, we assume the other variables are held constant. Take a look at the 3D graph of $f(x, y) = 2x^3 + 3y^3$ in Figure 1-8, and you will see we have slopes in two directions for two variables.

Let's take the function $f(x, y) = 2x^3 + 3y^3$. The x and y variable each get their own derivatives $\frac{d}{dx}$ and $\frac{d}{dy}$. These represent the slope values with respect to each variable on a multidimensional surface. We technically call these "slopes" *gradients* when dealing with multiple dimensions. These are the derivatives for x and y, followed by the SymPy code to calculate those derivatives:

$$f(x, y) = 2x^3 + 3y^3$$

$$\frac{d}{dx}2x^3 + 3y^3 = 6x^2$$

$$\frac{d}{dy}2x^3 + 3y^3 = 9y^2$$

Example 1-21 and Figure 1-8 show how we calculate the partial derivatives for x and y, respectively, with SymPy.

Example 1-21. Calculating partial derivatives with SymPy

```
from sympy import *
from sympy.plotting import plot3d

# Declare x and y to SymPy
x,y = symbols('x y')

# Now just use Python syntax to declare function
f = 2*x**3 + 3*y**3

# Calculate the partial derivatives for x and y
dx_f = diff(f, x)
```

```
dy_f = diff(f, y)

print(dx_f) # prints 6*x**2
print(dy_f) # prints 9*y**2

# plot the function
plot3d(f)
```

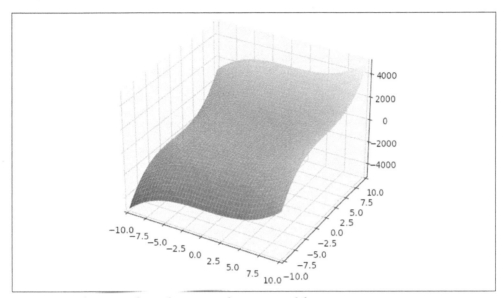

Figure 1-8. Plotting a three-dimensional exponential function

So for (x,y) values $(1,2)$, the slope with respect to x is $6(1) = 6$ and the slope with respect to y is $9(2)^2 = 36$.

Using Limits to Calculate Derivatives

Want to see where limits come into play calculating derivatives? If you are feeling good about what we learned so far, proceed! If you are still digesting, maybe consider coming back to this sidebar later.

SymPy allows us to do some interesting explorations about math. Take our function $f(x) = x^2$; we approximated a slope for $x = 2$ by drawing a line through a close neighboring point $x = 2.0001$ by adding a step 0.0001. Why not use a limit to forever decrease that step s and see what slope it approaches?

$$\lim_{s \to 0} \frac{(x + s)^2 - x^2}{(x + s) - x}$$

In our example, we are interested in the slope where $x = 2$ so let's substitute that:

$$\lim_{s \to 0} \frac{(2 + s)^2 - 2^2}{(2 + s) - 2} = 4$$

By forever approaching a step size s to 0 but never reaching it (remember the neighboring point cannot touch the point at $x = 2$, else we have no line!), we can use a limit to see we converge on a slope of 4 as shown in Example 1-22.

Example 1-22. Using limits to calculate a slope

```
from sympy import *

# "x" and step size "s"
x, s = symbols('x s')

# declare function
f = x**2

# slope between two points with gap "s"
# substitute into rise-over-run formula
slope_f = (f.subs(x, x + s) - f) / ((x+s) - x)

# substitute 2 for x
slope_2 = slope_f.subs(x, 2)

# calculate slope at x = 2
# infinitely approach step size _s_ to 0
result = limit(slope_2, s, 0)

print(result) # 4
```

Now what if we do not assign a specific value to x and leave it alone? What happens if we decrease our step size s infinitely toward 0? Let's look in Example 1-23.

Example 1-23. Using limits to calculate a derivative

```
from sympy import *

# "x" and step size "s"
x, s = symbols('x s')

# declare function
f = x**2

# slope between two points with gap "s"
# substitute into rise-over-run formula
slope_f = (f.subs(x, x + s) - f) / ((x+s) - x)

# calculate derivative function
```

```
# infinitely approach step size +s+ to 0
result = limit(slope_f, s, 0)

print(result) # 2x
```

That gave us our derivative function 2x. SymPy was smart enough to figure out to never let our step size reach 0 but forever approach 0. This converges $f(x) = x^2$ to reach its derivative counterpart $2x$.

The Chain Rule

In Chapter 7 when we build a neural network, we are going to need a special math trick called the chain rule. When we compose the neural network layers, we will have to untangle the derivatives from each layer. But for now let's learn the chain rule with a simple algebraic example. Let's say you are given two functions:

$$y = x^2 + 1$$
$$z = y^3 - 2$$

Notice that these two functions are linked, because the y is the output variable in the first function but is the input variable in the second. This means we can substitute the first function y into the second function z like this:

$$z = \left(x^2 + 1\right)^3 - 2$$

So what is the derivative for z with respect to x? We already have the substitution expressing z in terms of x. Let's use SymPy to calculate that in Example 1-24.

Example 1-24. Finding the derivative of z with respect to x

```
from sympy import *

x = symbols('x')

z = (x**2 + 1)**3 - 2
dz_dx = diff(z, x)
print(dz_dx)

# 6*x*(x**2 + 1)**2
```

So our derivative for z with respect to x is $6x\left(x^2 + 1\right)^2$:

$$\frac{dz}{dx}\left(\left(x^2+1\right)^3-2\right)$$
$$= 6x\left(x^2+1\right)^2$$

But look at this. Let's start over and take a different approach. If we take the derivatives of the y and z functions separately, and then multiply them together, this also produces the derivative of z with respect to x! Let's try it:

$$\frac{dy}{dx}\left(x^2+1\right) = 2x$$

$$\frac{dz}{dy}\left(y^3-2\right) = 3y^2$$

$$\frac{dz}{dx} = (2x)\left(3y^2\right) = 6xy^2$$

All right, $6xy^2$ may not look like $6x\left(x^2+1\right)^2$, but that's only because we have not substituted the y function yet. Do that so the entire $\frac{dz}{dx}$ derivative is expressed in terms of x without y.

$$\frac{dz}{dx} = 6xy^2 = 6x\left(x^2+1\right)^2$$

Now we see we got the same derivative function $6x\left(x^2+1\right)^2$!

This is the *chain rule*, which says that for a given function y (with input variable x) composed into another function z (with input variable y), we can find the derivative of z with respect to x by multiplying the two respective derivatives together:

$$\frac{dz}{dx} = \frac{dz}{dy} \times \frac{dy}{dx}$$

Example 1-25 shows the SymPy code that makes this comparison, showing the derivative from the chain rule is equal to the derivative of the substituted function.

Example 1-25. Calculating the derivative dz/dx with and without the chain rule, but still getting the same answer

```
from sympy import *

x, y = symbols('x y')


# derivative for first function
# need to underscore y to prevent variable clash
_y = x**2 + 1
dy_dx = diff(_y)

# derivative for second function
z = y**3 - 2
dz_dy = diff(z)

# Calculate derivative with and without
# chain rule, substitute y function
dz_dx_chain = (dy_dx * dz_dy).subs(y, _y)
dz_dx_no_chain = diff(z.subs(y, _y))

# Prove chain rule by showing both are equal
print(dz_dx_chain) # 6*x*(x**2 + 1)**2
print(dz_dx_no_chain) # 6*x*(x**2 + 1)**2
```

The chain rule is a key part of training a neural network with the proper weights and biases. Rather than untangle the derivative of each node in a nested onion fashion, we can multiply the derivatives across each node instead, which is mathematically a lot easier.

Integrals

The opposite of a derivative is an *integral*, which finds the area under the curve for a given range. In Chapters 2 and 3, we will be finding the areas under probability distributions. Although we will not use integrals directly, and instead will use cumulative density functions that are already integrated, it is good to be aware of how integrals find areas under curves. Appendix A contains examples of using this approach on probability distributions.

I want to take an intuitive approach for learning integrals called the Riemann Sums, one that flexibly adapts to any continuous function. First, let's point out that finding the area for a range under a straight line is easy. Let's say I have a function $f(x) = 2x$ and I want to find the area under the line between 0 and 1, as shaded in Figure 1-9.

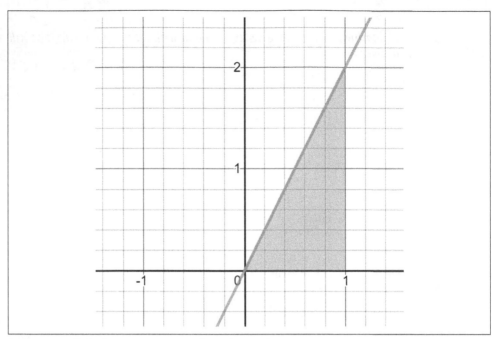

Figure 1-9. Calculating an area under a linear function

Notice that I am finding the area bounded between the line and the x-axis, and in the x range 0.0 to 1.0. If you recall basic geometry formulas, the area A for a triangle is $A = \frac{1}{2}bh$ where b is the length of the base and h is the height. We can visually spot that $b = 1$ and $h = 2$. So plugging into the formula, we get for our area 1.0 as calculated here:

$$A = \frac{1}{2}bh$$

$$A = \frac{1}{2} * 1 * 2$$

$$A = 1$$

That was not bad, right? But let's look at a function that is difficult to find the area under: $f(x) = x^2 + 1$. What is the area between 0 and 1 as shaded in Figure 1-10?

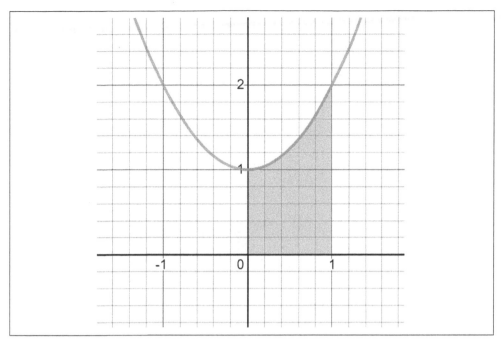

Figure 1-10. Calculating area under nonlinear functions is less straightforward

Again we are interested in the area below the curve and above the x-axis, only within the x range between 0 and 1. The curviness here does not give us a clean geometric formula to find the area, but here is a clever little hack you can do.

What if we packed five rectangles of equal length under the curve as shown in Figure 1-11, where the height of each one extends from the x-axis to where the midpoint touches the curve?

The area of a rectangle is A = length × width, so we could easily sum the areas of the rectangles. Would that give us a good approximation of the area under the curve? What if we packed 100 rectangles? 1,000? 100,000? As we increase the number of rectangles while decreasing their width, would we not get closer to the area under the curve? Yes we would, and it is yet another case where we increase/decrease something toward infinity to approach an actual value.

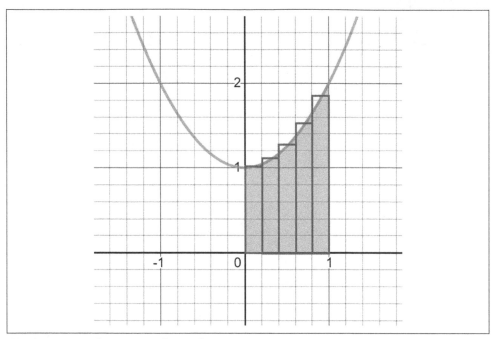

Figure 1-11. Packing rectangles under a curve to approximate area

Let's try it out in Python. First we need a function that approximates an integral that we will call `approximate_integral()`. The arguments a and b will specify the min and max of the *x* range, respectively. n will be the number of rectangles to pack, and f will be the function we are integrating. We implement the function in Example 1-26, and then use it to integrate our function $f(x) = x^2 + 1$ with five rectangles, between 0.0 and 1.0.

Example 1-26. An integral approximation in Python

```
def approximate_integral(a, b, n, f):
    delta_x = (b - a) / n
    total_sum = 0

    for i in range(1, n + 1):
        midpoint = 0.5 * (2 * a + delta_x * (2 * i - 1))
        total_sum += f(midpoint)

    return total_sum * delta_x

def my_function(x):
    return x**2 + 1

area = approximate_integral(a=0, b=1, n=5, f=my_function)
```

```
print(area) # prints 1.33
```

So we get an area of 1.33. What happens if we use 1,000 rectangles? Let's try it out in Example 1-27.

Example 1-27. Another integral approximation in Python

```
area = approximate_integral(a=0, b=1, n=1000, f=my_function)

print(area) # prints 1.333333250000001
```

OK, we are getting some more precision here, and getting some more decimal places. What about one million rectangles as shown in Example 1-28?

Example 1-28. Yet another integral approximation in Python

```
area = approximate_integral(a=0, b=1, n=1_000_000, f=my_function)

print(area) # prints 1.3333333333332733
```

OK, I think we are getting a diminishing return here and converging on the value $1.\overline{333}$ where the ".333" part is forever recurring. If this were a rational number, it is likely $4/3 = 1.\overline{333}$. As we increase the number of rectangles, the approximation starts to reach its limit at smaller and smaller decimals.

Now that we have some intuition on what we are trying to achieve and why, let's do a more exact approach with SymPy, which happens to support rational numbers, in Example 1-29.

Example 1-29. Using SymPy to perform integration

```
from sympy import *

# Declare 'x' to SymPy
x = symbols('x')

# Now just use Python syntax to declare function
f = x**2 + 1

# Calculate the integral of the function with respect to x
# for the area between x = 0 and 1
area = integrate(f, (x, 0, 1))

print(area) # prints 4/3
```

Cool! So the area actually is 4/3, which is what our previous method converged on. Unfortunately, plain Python (and many programming languages) only support decimals, but computer algebra systems like SymPy give us exact rational numbers. We will be using integrals to find areas under curves in Chapters 2 and 3, although we will have scikit-learn do the work for us.

Calculating Integrals with Limits

For the super curious, here is how we calculate definite integrals using limits in SymPy. Please! Skip or come back to this sidebar later if you have enough to digest. But if you are feeling good and want to deep dive into how integrals are derived using limits, proceed!

The main idea follows much of what we did earlier: pack rectangles under a curve and make them infinitely smaller until we approach the exact area. But of course, the rectangles cannot have a width of 0...they just have to keep getting closer to 0 without ever reaching 0. It is another case of using limits.

Khan Academy has a great article (*https://oreil.ly/sBmCy*) explaining how to use limits for Riemann Sums but here is how we do it in SymPy as shown in Example 1-30.

Example 1-30. Using limits to calculate integrals

```python
from sympy import *

# Declare variables to SymPy
x, i, n = symbols('x i n')

# Declare function and range
f = x**2 + 1
lower, upper = 0, 1

# Calculate width and each rectangle height at index "i"
delta_x = ((upper - lower) / n)
x_i = (lower + delta_x * i)
fx_i = f.subs(x, x_i)

# Iterate all "n" rectangles and sum their areas
n_rectangles = Sum(delta_x * fx_i, (i, 1, n)).doit()

# Calculate the area by approaching the number
# of rectangles "n" to infinity
area = limit(n_rectangles, n, oo)

print(area) # prints 4/3
```

Here we determine the length of each rectangle delta_x and the start of each rectangle x_i where i is the index of each rectangle. fx_i is the height of each rectangle at index i. We declare n number of rectangles and sum their areas delta_x * fx_i, but we have no area value yet because we have not committed a number for n. Instead we approach n toward infinity to see what area we converge on, and you should get 4/3!

Conclusion

In this chapter we covered some foundations we will use for the rest of this book. From number theory to logarithms and calculus integrals, we highlighted some important mathematical concepts relevant to data science, machine learning, and analytics. You may have questions about why these concepts are useful. That will come next!

Before we move on to discuss probability, take a little time to skim these concepts one more time and then do the following exercises. You can always revisit this chapter as you progress through this book and refresh as necessary when you start applying these mathematical ideas.

Exercises

1. Is the value 62.6738 rational or irrational? Why or why not?

2. Evaluate the expression: $10^7 10^{-5}$

3. Evaluate the expression: $81^{\frac{1}{2}}$

4. Evaluate the expression: $25^{\frac{3}{2}}$

5. Assuming no payments are made, how much would a $1,000 loan be worth at 5% interest compounded monthly after 3 years?

6. Assuming no payments are made, how much would a $1,000 loan be worth at 5% interest compounded continuously after 3 years?

7. For the function $f(x) = 3x^2 + 1$ what is the slope at $x = 3$?

8. For the function $f(x) = 3x^2 + 1$ what is the area under the curve for x between 0 and 2?

Answers are in Appendix B.

## CHAPTER 2

Probability

When you think of probability, what images come to mind? Perhaps you think of gambling-related examples, like the probability of winning the lottery or getting a pair with two dice. Maybe it is predicting stock performance, the outcome of a political election, or whether your flight will arrive on time. Our world is full of uncertainties we want to measure.

Maybe that is the word we should focus on: uncertainty. How do we measure something that we are uncertain about?

In the end, probability is the theoretical study of measuring certainty that an event will happen. It is a foundational discipline for statistics, hypothesis testing, machine learning, and other topics in this book. A lot of folks take probability for granted and assume they understand it. However, it is more nuanced and complicated than most people think. While the theorems and ideas of probability are mathematically sound, it gets more complex when we introduce data and venture into statistics. We will cover that in Chapter 3 on statistics and hypothesis testing.

In this chapter, we will discuss what probability is. Then we will cover probability math concepts, Bayes' Theorem, the binomial distribution, and the beta distribution.

Understanding Probability

Probability is how strongly we believe an event will happen, often expressed as a percentage. Here are some questions that might warrant a probability for an answer:

- How likely will I get 7 heads in 10 fair coin flips?
- What are my chances of winning an election?
- Will my flight be late?
- How certain am I that a product is defective?

The most popular way to express probability is as a percentage, as in "There is a 70% chance my flight will be late." We will call this probability $P(X)$, where X is the event of interest. As you work with probabilities, though, you will more likely see it expressed as a decimal (in this case .70), which must be between 0.0 and 1.0:

$$P(X) = .70$$

Likelihood is similar to probability, and it is easy to confuse the two (many dictionaries do as well). You can get away with using "probability" and "likelihood" interchangeably in everyday conversation. However, we should pin down these differences. Probability is about quantifying predictions of events yet to happen, whereas likelihood is measuring the frequency of events that already occurred. In statistics and machine learning, we often use likelihood (the past) in the form of data to predict probability (the future).

It is important to note that a probability of an event happening must be strictly between 0% and 100%, or 0.0 and 1.0. Logically, this means the probability of an event *not* happening is calculated by subtracting the probability of the event from 1.0:

$$P(X) = .70$$
$$P(\text{not } X) = 1 - .70 = .30$$

This is another distinction between probability and likelihood. Probabilities of all possible mutually exclusive outcomes for an event (meaning only one outcome can occur, not multiple) must sum to 1.0 or 100%. Likelihoods, however, are not subject to this rule.

Alternatively, probability can be expressed as an *odds* $O(X)$ such as 7:3, 7/3, or $2.\overline{333}$.

To turn an odds $O(X)$ into a proportional probability $P(X)$, use this formula:

$$P(X) = \frac{O(X)}{1 + O(X)}$$

So if I have an odds 7/3, I can convert it into a proportional probability like this:

$$P(X) = \frac{O(X)}{1 + O(X)}$$

$$P(X) = \frac{\frac{7}{3}}{1 + \frac{7}{3}}$$

$$P(X) = .7$$

Conversely, you can turn a probability into an odds simply by dividing the probability of the event occurring by the probability it will not occur:

$$O(X) = \frac{P(X)}{1 - P(X)}$$

$$O(X) = \frac{.70}{1 - .70}$$

$$O(X) = \frac{7}{3}$$

Odds Are Useful!

While many people feel more comfortable expressing probabilities as percentages or proportions, odds can be a helpful tool. If I have an odds of 2.0, that means I feel an event is two times more likely to happen than not to happen. That can be more intuitive to describe a belief than a percentage of $66.\overline{666}$%. For this reason, odds are helpful quantifying subjective beliefs especially in a gambling/betting context. It plays a role in Bayesian statistics (including the Bayes Factor) as well as logistic regressions with the log-odds, which we will cover in Chapter 6.

Probability Versus Statistics

Sometimes people use the terms *probability* and *statistics* interchangeably, and while it is understandable to conflate the two disciplines, they do have distinctions. *Probability* is purely theoretical of how likely an event is to happen and does not require data. *Statistics*, on the other hand, cannot exist without data and uses it to discover probability and provides tools to describe data.

Think of predicting the outcome of rolling a 4 on a die (that's the singular of dice). Approaching the problem with a pure probability mindset, one simply says there are six sides on a die. We assume each side is equally likely, so the probability of getting a 4 is 1/6, or 16.666%.

However, a zealous statistician might say, "No! We need to roll the die to get data. If we can get 30 rolls or more, and the more rolls we do the better, only then will we have data to determine the probability of getting a 4." This approach may seem silly if we assume the die is fair, but what if it's not? If that's the case, collecting data is the only way to discover the probability of rolling a 4. We will talk about hypothesis testing in Chapter 3.

Probability Math

When we work with a single probability of an event $P(X)$, known as a *marginal probability*, the idea is fairly straightforward, as discussed previously. But when we start combining probabilities of different events, it gets a little less intuitive.

Joint Probabilities

Let's say you have a fair coin and a fair six-sided die. You want to find the probability of flipping a heads and rolling a 6 on the coin and die, respectively. These are two separate probabilities of two separate events, but we want to find the probability that both events will occur together. This is known as a *joint probability*.

Think of a joint probability as an AND operator. I want to find the probability of flipping a heads AND rolling a 6. We want both events to happen together, so how do we calculate this probability?

There are two sides on a coin and six sides on the die, so the probability of heads is 1/2 and the probability of six is 1/6. The probability of both events occurring (assuming they are independent, more on this later!) is simply multiplying the two together:

$$P(A \text{ AND } B) = P(A) \times P(B)$$

$$P(\text{heads}) = \frac{1}{2}$$

$$P(6) = \frac{1}{6}$$

$$P(\text{heads AND } 6) = \frac{1}{2} \times \frac{1}{6} = \frac{1}{12} = .08\overline{333}$$

Easy enough, but why is this the case? A lot of probability rules can be discovered by generating all possible combinations of events, which comes from an area of discrete math known as permutations and combinations. For this case, generate every possible outcome between the coin and die, pairing heads (H) and tails (T) with the numbers 1 through 6. Note I put asterisks "*" around the outcome of interest where we get heads and a 6:

H1 H2 H3 H4 H5 *H6* T1 T2 T3 T4 T5 T6

Notice there are 12 possible outcomes when flipping our coin and rolling our die. The only one that is of interest to us is "H6," getting a heads and a 6. So because there is only one outcome that satisfies our condition, and there are 12 possible outcomes, the probability of getting a heads and a 6 is 1/12.

Rather than generate all possible combinations and counting the ones of interest to us, we can again use the multiplication as a shortcut to find the joint probability. This is known as the *product rule*:

$$P(A \text{ AND } B) = P(A) \times P(B)$$

$$P(\text{heads AND } 6) = \frac{1}{2} \times \frac{1}{6} = \frac{1}{12} = .08\overline{333}$$

Union Probabilities

We discussed joint probabilities, which is the probability of two or more events occurring simultaneously. But what about the probability of getting event A or B? When we deal with OR operations with probabilities, this is known as a *union probability*.

Let's start with *mutually exclusive* events, which are events that cannot occur simultaneously. For example, if I roll one die I cannot simultaneously get a 4 and a 6. I can only get one outcome. Getting the union probability for these cases is easy. I simply add them together. If I want to find the probability of getting a 4 or 6 on a die roll, it is going to be 2/6 = 1/3:

$$P(4) = \frac{1}{6}$$

$$P(6) = \frac{1}{6}$$

$$P(4 \text{ OR } 6) = \frac{1}{6} + \frac{1}{6} = \frac{1}{3}$$

But what about *nonmutually exclusive* events, which are events that can occur simultaneously? Let's go back to the coin flip and die roll example. What is the probability of getting a heads OR a 6? Before you are tempted to add those probabilities, let's generate all possible outcomes again and highlight the ones we are interested in:

H1 *H2* *H3* *H4* *H5* *H6* T1 T2 T3 T4 T5 *T6*

Here we are interested in all the heads outcomes as well as the 6 outcomes. If we proportion the 7 out of 12 outcomes we are interested in, 7/12, we get a correct probability of .58$\overline{333}$.

But what happens if we add the probabilities of heads and 6 together? We get a different (and wrong!) answer of $.\overline{666}$:

$$P(heads) = \frac{1}{2}$$

$$P(6) = \frac{1}{6}$$

$$P(heads \text{ OR } 6) = \frac{1}{2} + \frac{1}{6} = \frac{4}{6} = .\overline{666}$$

Why is that? Study the combinations of coin flip and die outcomes again and see if you can find something fishy. Notice when we add the probabilities, we double-count the probability of getting a 6 in both "H6" and "T6"! If this is not clear, try finding the probability of getting heads or a die roll of 1 through 5:

$$P(heads) = \frac{1}{2}$$

$$P(1 \text{ through } 5) = \frac{5}{6}$$

$$P(heads \text{ OR } 1 \text{ through } 5) = \frac{1}{2} + \frac{5}{6} = \frac{8}{6} = 1.\overline{333}$$

We get a probability of 133.333%, which is definitely not correct because a probability must be no more than 100% or 1.0. The problem again is we are double-counting outcomes.

If you ponder long enough, you may realize the logical way to remove double-counting in a union probability is to subtract the joint probability. This is known as the *sum rule of probability* and ensures every joint event is counted only once:

$$P(A \text{ OR } B) = P(A) + P(B) - P(A \text{ AND } B)$$
$$P(A \text{ OR } B) = P(A) + P(B) - P(A) \times P(B)$$

So going back to our example of calculating the probability of a heads or a 6, we need to subtract the joint probability of getting a heads or a 6 from the union probability:

$$P(heads) = \frac{1}{2}$$

$$P(6) = \frac{1}{6}$$

$$P(A \text{ OR } B) = P(A) + P(B) - P(A) \times P(B)$$

$$P(heads \text{ OR } 6) = \frac{1}{2} + \frac{1}{6} - \left(\frac{1}{2} \times \frac{1}{6}\right) = .58\overline{333}$$

Note that this formula also applies to mutually exclusive events. If the events are mutually exclusive where only one outcome A or B is allowed but not both, then the joint probability P(A AND B) is going to be 0, and therefore remove itself from the formula. You are then left with simply summing the events as we did earlier.

In summary, when you have a union probability between two or more events that are not mutually exclusive, be sure to subtract the joint probability so no probabilities are double-counted.

Conditional Probability and Bayes' Theorem

A probability topic that easily confuses people is the concept of *conditional probability*, which is the probability of an event A occurring given event B has occurred. It is typically expressed as $P(A\ GIVEN\ B)$ or $P(A|B)$.

Let's say a study makes a claim that 85% of cancer patients drank coffee. How do you react to this claim? Does this alarm you and make you want to abandon your favorite morning drink? Let's first define this as a conditional probability $P(Coffee\ given\ Cancer)$ or $P(Coffee|Cancer)$. This represents a probability of people who drink coffee given they have cancer.

Within the United States, let's compare this to the percentage of people diagnosed with cancer (0.5% according to *cancer.gov*) and the percentage of people who drink coffee (65% according to *statista.com*):

$P(Coffee) = .65$
$P(Cancer) = .005$
$P(Coffee|Cancer) = .85$

Hmmmm...study these numbers for a moment and ask whether coffee is really the problem here. Notice again that only 0.5% of the population has cancer at any given time. However 65% of the population drinks coffee regularly. If coffee contributes to cancer, should we not have much higher cancer numbers than 0.5%? Would it not be closer to 65%?

This is the sneaky thing about proportional numbers. They may seem significant without any given context, and media headlines can certainly exploit this for clicks: "New Study Reveals 85% of Cancer Patients Drink Coffee" it might read. Of course, this is silly because we have taken a common attribute (drinking coffee) and associated it with an uncommon one (having cancer).

The reason people can be so easily confused by conditional probabilities is because the direction of the condition matters, and the two conditions are conflated as somehow being equal. The "probability of having cancer given you are a coffee drinker" is different from the "probability of being a coffee drinker given you have cancer." To put it simply: few coffee drinkers have cancer, but many cancer patients drink coffee.

If we are interested in studying whether coffee contributes to cancer, we really are interested in the first conditional probability: the probability of someone having cancer given they are a coffee drinker.

$P(\text{Coffee}|\text{Cancer}) = .85$

$P(\text{Cancer}|\text{Coffee}) = ?$

How do we flip the condition? There's a powerful little formula called *Bayes' Theorem*, and we can use it to flip conditional probabilities:

$$P(A|B) = \frac{P(B|A)P(A)}{P(B)}$$

If we plug the information we already have into this formula, we can solve for the probability someone has cancer given they drink coffee:

$$P(A|B) = \frac{P(B|A) * P(A)}{P(B)}$$

$$P(\text{Cancer}|\text{Coffee}) = \frac{P(\text{Coffee}|\text{Cancer}) * P(\text{Cancer})}{P(\text{Coffee})}$$

$$P(\text{Cancer}|\text{Coffee}) = \frac{.85 * .005}{.65} = .0065$$

If you want to calculate this in Python, check out Example 2-1.

Example 2-1. Using Bayes' Theorem in Python

```
p_coffee_drinker = .65
p_cancer = .005
p_coffee_drinker_given_cancer = .85

p_cancer_given_coffee_drinker = p_coffee_drinker_given_cancer *
    p_cancer / p_coffee_drinker

# prints 0.006538461538461539
print(p_cancer_given_coffee_drinker)
```

So the probability someone has cancer given they are a coffee drinker is only 0.65%! This number is very different from the probability someone is a coffee drinker given they have cancer, which is 85%. Now do you see why the direction of the condition matters? Bayes' Theorem is helpful for this reason. It can also be used to chain several conditional probabilities together to keep updating our beliefs based on new information.

What Defines a "Coffee Drinker"?

Note that I could have accounted for other variables here, in particular what qualifies someone as a "coffee drinker." If someone drinks coffee once a month, as opposed to someone who drinks coffee every day, should I equally qualify both as "coffee drinkers"? What about the person who started drinking coffee a month ago as opposed to someone who drank coffee for 20 years? How often and how long do people have to drink coffee before they meet the threshold of being a "coffee drinker" in this cancer study?

These are important questions to consider, and they show why data rarely tells the whole story. If someone gives you a spreadsheet of patients with a simple "YES/NO" flag on whether they are a coffee drinker, that threshold needs to be defined! Or we need a more weightful metric like "number of coffee drinks consumed in the last three years." I kept this thought experiment simple and didn't define how someone qualifies as a "coffee drinker," but be aware that out in the field, it is always a good idea to pull the threads on the data. We will discuss this more in Chapter 3.

If you want to explore the intuition behind Bayes' Theorem more deeply, turn to Appendix A. For now just know it helps us flip a conditional probability. Let's talk about how conditional probabilities interact with joint and union operations next.

Naive Bayes

Bayes' Theorem plays a central role in a common machine learning algorithm called Naive Bayes. Joel Grus covers it in his book *Data Science from Scratch* (O'Reilly).

Joint and Union Conditional Probabilities

Let's revisit joint probabilities and how they interact with conditional probabilities. I want to find the probability somebody is a coffee drinker AND they have cancer. Should I multiply $P(\text{Coffee})$ and $P(\text{Cancer})$? Or should I use $P(\text{Coffee}|\text{Cancer})$ in place of $P(\text{Coffee})$ if it is available? Which one do I use?

Option 1:
$P(\text{Coffee}) \times P(\text{Cancer}) = .65 \times .005 = .00325$

Option 2:
$P(\text{Coffee}|\text{Cancer}) \times P(\text{Cancer}) = .85 \times .005 = .00425$

If we already have established our probability applies only to people with cancer, does it not make sense to use $P(\text{Coffee}|\text{Cancer})$ instead of $P(\text{Coffee})$? One is more specific and applies to a condition that's already been established. So we should use $P(\text{Coffee}|\text{Cancer})$ as $P(\text{Cancer})$ is already part of our joint probability. This means the probability of someone having cancer and being a coffee drinker is 0.425%:

$$P(\text{Coffee and Cancer}) = P(\text{Coffee}|\text{Cancer}) \times P(\text{Cancer}) = .85 \times .005 = .00425$$

This joint probability also applies in the other direction. I can find the probability of someone being a coffee drinker and having cancer by multiplying $P(\text{Cancer}|\text{Coffee})$ and $P(\text{Coffee})$. As you can observe, I arrive at the same answer:

$$P(\text{Cancer}|\text{Coffee}) \times P(\text{Coffee}) = .0065 \times .65 = .00425$$

If we did not have any conditional probabilities available, then the best we can do is multiply $P(\text{Coffee Drinker})$ and $P(\text{Cancer})$ as shown here:

$$P(\text{Coffee Drinker}) \times P(\text{Cancer}) = .65 \times .005 = .00325$$

Now think about this: if event A has no impact on event B, then what does that mean for conditional probability $P(B|A)$? That means $P(B|A) = P(B)$, meaning event A occurring makes no difference to how likely event B is to occur. Therefore we can update our joint probability formula, regardless if the two events are dependent, to be:

$$P(A \text{ AND } B) = P(B) \times P(A|B)$$

And finally let's talk about unions and conditional probability. If I wanted to calculate the probability of A or B occurring, but A may affect the probability of B, we update our sum rule like this:

$$P(A \text{ OR } B) = P(A) + P(B) - P(A|B) \times P(B)$$

As a reminder, this applies to mutually exclusive events as well. The sum rule $P(A|B) \times P(B)$ would yield 0 if the events A and B cannot happen simultaneously.

Binomial Distribution

For the remainder of this chapter, we are going to learn two probability distributions: the binomial and beta distributions. While we will not be using these for the rest of the book, they are useful tools in themselves and fundamental to understanding how events occur given a number of trials. They will also be a good segue to understanding probability distributions that we will use heavily in Chapter 3. Let's explore a use case that could occur in a real-world scenario.

Let's say you are working on a new turbine jet engine and you ran 10 tests. The outcomes yielded eight successes and two failures:

✓ ✓ ✓ ✓ ✓ ✗ ✓ ✗ ✓ ✓

You were hoping to get a 90% success rate, but based on this data you conclude that your tests have failed with only 80% success. Each test is time-consuming and expensive, so you decide it is time to go back to the drawing board to reengineer the design.

However, one of your engineers insists there should be more tests. "The only way we will know for sure is to run more tests," she argues. "What if more tests yield 90% or greater success? After all, if you flip a coin 10 times and get 8 heads, it does not mean the coin is fixed at 80%."

You briefly consider the engineer's argument and realize she has a point. Even a fair coin flip will not always have an equally split outcome, especially with only 10 flips. You are most likely to get five heads but you can also get three, four, six, or seven heads. You could even get 10 heads, although this is extremely unlikely. So how do you determine the likelihood of 80% success assuming the underlying probability is 90%?

One tool that might be relevant here is the *binomial distribution*, which measures how likely k successes can happen out of n trials given p probability.

Visually, a binomial distribution looks like Figure 2-1.

Here, we see the probability of k successes for each bar out of 10 trials. This binomial distribution assumes a probability p of 90%, meaning there is a .90 (or 90%) chance for a success to occur. If this is true, that means there is a .1937 probability we would get 8 successes out of 10 trials. The probability of getting 1 success out of 10 trials is extremely unlikely, .000000008999, hence why the bar is not even visible.

We can also calculate the probability of eight or fewer successes by adding up bars for eight or fewer successes. This would give us .2639 probability of eight or fewer successes.

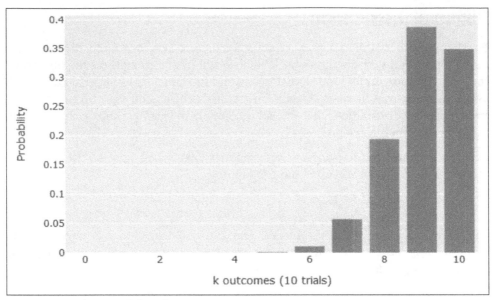

Figure 2-1. A binomial distribution

So how do we implement the binomial distribution? We can do it from scratch relatively easily (as shared in Appendix A), or we can use libraries like SciPy. Example 2-2 shows how we use SciPy's `binom.pmf()` function (*PMF* stands for "probability mass function") to print all 11 probabilities for our binomial distribution from 0 to 10 successes.

Example 2-2. Using SciPy for the binomial distribution

```
from scipy.stats import binom

n = 10
p = 0.9

for k in range(n + 1):
    probability = binom.pmf(k, n, p)
    print("{0} - {1}".format(k, probability))


# OUTPUT:

# 0 - 9.99999999999996e-11
# 1 - 8.999999999999996e-09
# 2 - 3.644999999999996e-07
# 3 - 8.748000000000003e-06
# 4 - 0.0001377809999999999
# 5 - 0.0014880347999999988
# 6 - 0.011160260999999996
```

```
# 7 - 0.05739562800000001
# 8 - 0.19371024449999993
# 9 - 0.38742048900000037
# 10 - 0.34867844010000004
```

As you can see, we provide *n* as the number of trials, *p* as the probability of success for each trial, and *k* as the number of successes we want to look up the probability for. We iterate each number of successes *k* with the corresponding probability we would see that many successes. As we can see in the output, the most likely number of successes is nine.

But if we add up the probability of eight or fewer successes, we would get .2639. This means there is a 26.39% chance we would see eight or fewer successes even if the underlying success rate is 90%. So maybe the engineer is right: 26.39% chance is not nothing and certainly possible.

However, we did make an assumption here in our model, which we will discuss next with the beta distribution.

Binomial Distribution from Scratch

Turn to Appendix A to learn how to build the binomial distribution from scratch without SciPy.

Beta Distribution

What did I assume with my engine-test model using the binomial distribution? Is there a parameter I assumed to be true and then built my entire model around it? Think carefully and read on.

What might be problematic about my binomial distribution is I *assumed* the underlying success rate is 90%. That's not to say my model is worthless. I just showed if the underlying success rate is 90%, there is a 26.39% chance I would see 8 or fewer successes with 10 trials. So the engineer is certainly not wrong that there could be an underlying success rate of 90%.

But let's flip the question and consider this: what if there are other underlying rates of success that would yield 8/10 successes besides 90%? Could we see 8/10 successes with an underlying 80% success rate? 70%? 30%? When we fix the 8/10 successes, can we explore the probabilities of probabilities?

Rather than create countless binomial distributions to answer this question, there is one tool that we can use. The *beta distribution* allows us to see the likelihood of different underlying probabilities for an event to occur given *alpha* successes and *beta* failures.

A chart of the beta distribution given eight successes and two failures is shown in Figure 2-2.

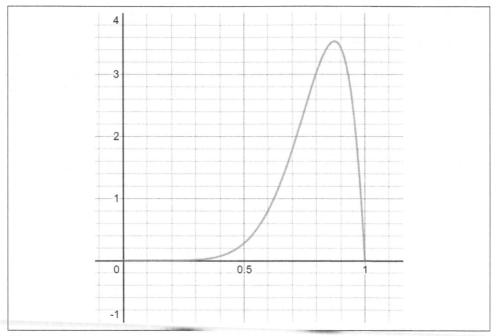

Figure 2-2. Beta distribution

Beta Distribution on Desmos

If you want to interact with the beta distribution, a Desmos graph is provided here (*https://oreil.ly/pN4Ep*).

Notice that the x-axis represents all underlying rates of success from 0.0 to 1.0 (0% to 100%), and the y-axis represents the likelihood of that probability given eight successes and two failures. In other words, the beta distribution allows us to see the probabilities of probabilities given 8/10 successes. Think of it as a meta-probability so take your time grasping this idea!

Notice also that the beta distribution is a continuous function, meaning it forms a continuous curve of decimal values (as opposed to the tidy and discrete integers in the binomial distribution). This is going to make the math with the beta distribution a bit harder, as a given density value on the y-axis is not a probability. We instead find probabilities using areas under the curve.

The beta distribution is a type of *probability distribution*, which means the area under the entire curve is 1.0, or 100%. To find a probability, we need to find the area within a range. For example, if we want to evaluate the probability 8/10 successes would yield 90% or higher success rate, we need to find the area between 0.9 and 1.0, which is .225, as shaded in Figure 2-3.

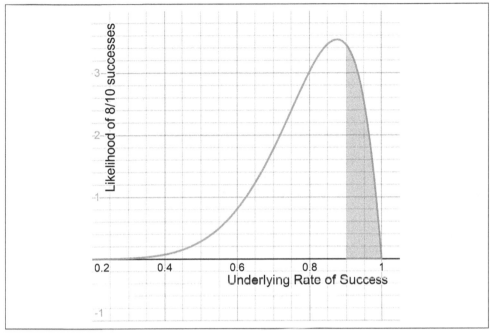

Figure 2-3. The area between 90% and 100%, which is 22.5%

As we did with the binomial distribution, we can use SciPy to implement the beta distribution. Every continuous probability distribution has a *cumulative density function (CDF)*, which calculates the area up to a given x-value. Let's say I wanted to calculate the area up to 90% (0.0 to 0.90) as shaded in Figure 2-4.

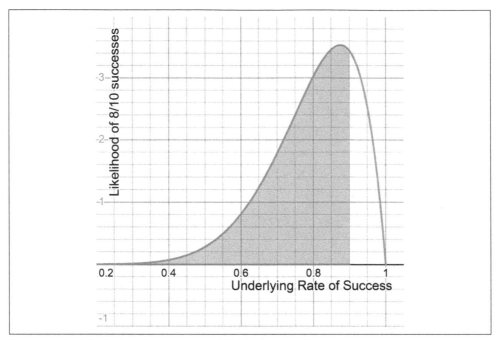

Figure 2-4. Calculating the area up to 90% (0.0 to 0.90)

It is easy enough to use SciPy with its beta.cdf() function, and the only parameters I need to provide are the x-value, the number of successes *a*, and the number of failures *b* as shown in Example 2-3.

Example 2-3. Beta distribution using SciPy

```
from scipy.stats import beta

a = 8
b = 2

p = beta.cdf(.90, a, b)

# 0.7748409780000001
print(p)
```

So according to our calculation, there is a 77.48% chance the underlying probability of success is 90% or less.

How do we calculate the probability of success being 90% or more as shaded in Figure 2-5?

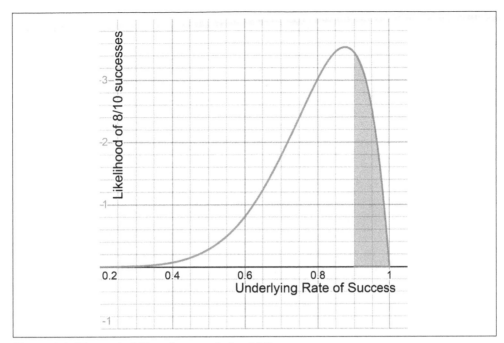

Figure 2-5. *The probability of success being 90% or more*

Our CDF calculates area only to the left of our boundary, not the right. Think about our rules of probability, and with a probability distribution the total area under the curve is 1.0. If we want to find the opposite probability of an event (greater than 0.90 as opposed to less than 0.90), just subtract the probability of being less than 0.90 from 1.0, and the remaining probability will capture being greater than 0.90. Figure 2-6 illustrates how we do this subtraction.

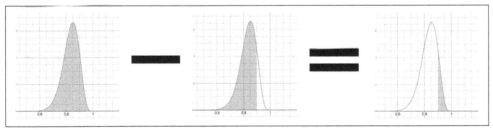

Figure 2-6. *Finding the probability of success being greater than 90%*

Example 2-4 shows how we calculate this subtraction operation in Python.

Example 2-4. Subtracting to get a right area in a beta distribution

```
from scipy.stats import beta

a = 8
b = 2

p = 1.0 - beta.cdf(.90, a, b)

# 0.22515902199999993
print(p)
```

This means that out of 8/10 successful engine tests, there is only a 22.5% chance the underlying success rate is 90% or greater. But there is about a 77.5% chance it is less than 90%. The odds are not in our favor here that our tests were successful, but we could gamble on that 22.5% chance with more tests if we are feeling lucky. If our CFO granted funding for 26 more tests resulting in 30 successes and 6 failures, our beta distribution would look like Figure 2-7.

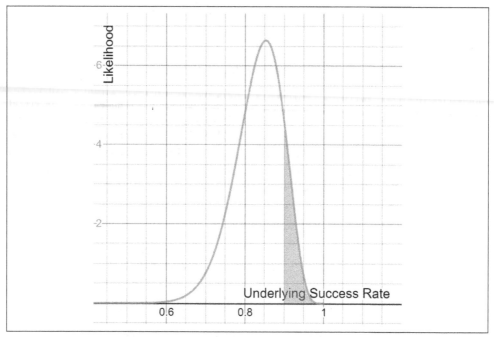

Figure 2-7. Beta distribution after 30 successes and 6 failures

Notice our distribution became narrower, thus becoming more confident that the underlying rate of success is in a smaller range. Unfortunately, our probability of meeting our 90% success rate minimum has decreased, going from 22.5% to 13.16% as shown in Example 2-5.

Example 2-5. A beta distribution with more trials

```
from scipy.stats import beta

a = 30
b = 6

p = 1.0 - beta.cdf(.90, a, b)

# 0.13163577484183708
print(p)
```

At this point, it might be a good idea to walk away and stop doing tests, unless you want to keep gambling against that 13.16% chance and hope the peak moves to the right.

Last but not least, how would we calculate an area in the middle? What if I want to find the probability my underlying rate of success is between 80% and 90% as shown in Figure 2-8?

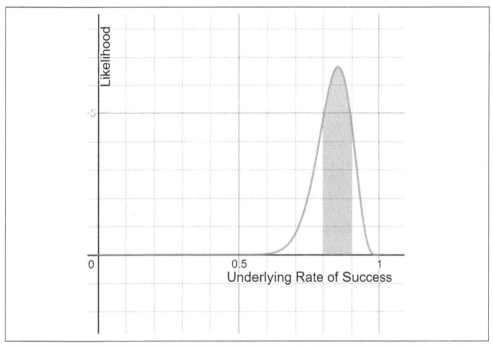

Figure 2-8. Probability the underlying rate of success is between 80% and 90%

Think carefully how you might approach this. What if we were to subtract the area behind .80 from the area behind .90 like in Figure 2-9?

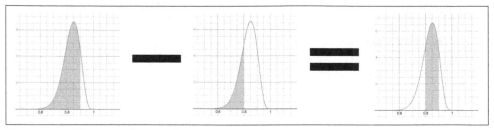

Figure 2-9. Obtaining the area between .80 and .90

Would that give us the area between .80 and .90? Yes it would, and it would yield an area of .3386 or 33.86% probability. Here is how we would calculate it in Python (Example 2-6).

Example 2-6. Beta distribution middle area using SciPy

```
from scipy.stats import beta

a = 8
b = 2

p = beta.cdf(.90, a, b) - beta.cdf(.80, a, b)

# 0.33863336200000010
print(p)
```

The beta distribution is a fascinating tool to measure the probability of an event occurring versus not occurring, based on a limited set of observations. It allows us to reason about probabilities of probabilities, and we can update it as we get new data. We can also use it for hypothesis testing, but we will put more emphasis on using the normal distribution and T-distribution for that purpose in Chapter 3.

Beta Distribution from Scratch

To learn how to implement the beta distribution from scratch, refer to Appendix A.

Conclusion

We covered a lot in this chapter! We not only discussed the fundamentals of probability, its logical operators, and Bayes' Theorem, but also introduced probability distributions, including the binomial and beta distributions. In the next chapter we will cover one of the more famous distributions, the normal distribution, and how it relates to hypothesis testing.

If you want to learn more about Bayesian probability and statistics, a great book is *Bayesian Statistics the Fun Way* by Will Kurt (No Starch Press). There are also interactive Katacoda scenarios available on the O'Reilly platform (*https://oreil.ly/OFbai*).

Exercises

1. There is a 30% chance of rain today, and a 40% chance your umbrella order will arrive on time. You are eager to walk in the rain today and cannot do so without either!

   What is the probability it will rain AND your umbrella will arrive?

2. There is a 30% chance of rain today, and a 40% chance your umbrella order will arrive on time.

   You will be able to run errands only if it does not rain or your umbrella arrives.

   What is the probability it will not rain OR your umbrella arrives?

3. There is a 30% chance of rain today, and a 40% chance your umbrella order will arrive on time.

   However, you found out if it rains there is only a 20% chance your umbrella will arrive on time.

   What is the probability it will rain AND your umbrella will arrive on time?

4. You have 137 passengers booked on a flight from Las Vegas to Dallas. However, it is Las Vegas on a Sunday morning and you estimate each passenger is 40% likely to not show up.

   You are trying to figure out how many seats to overbook so the plane does not fly empty.

   How likely is it at least 50 passengers will not show up?

5. You flipped a coin 10 times and got heads 8 times and tails 2 times.

   Do you think this coin has any good probability of being fair? Why or why not?

Answers are in Appendix B.

## CHAPTER 3

Descriptive and Inferential Statistics

Statistics is the practice of collecting and analyzing data to discover findings that are useful or predict what causes those findings to happen. Probability often plays a large role in statistics, as we use data to estimate how likely an event is to happen.

It may not always get credit, but statistics is the heart of many data-driven innovations. Machine learning in itself is a statistical tool, searching for possible hypotheses to correlate relationships between different variables in data. However there are a lot of blind sides in statistics, even for professional statisticians. We can easily get caught up in what the data says that we forget to ask where the data comes from. These concerns become all the more important as big data, data mining, and machine learning all accelerate the automation of statistical algorithms. Therefore, it is important to have a solid foundation in statistics and hypothesis testing so you do not treat these automations as black boxes.

In this section we will cover the fundamentals of statistics and hypothesis testing. Starting with descriptive statistics, we will learn common ways to summarize data. After that, we will venture into inferential statistics, where we try to uncover attributes of a population based on a sample.

What Is Data?

It may seem odd to define "data," something we all use and take for granted. But I think it needs to be done. Chances are if you asked any person what data is, they might answer to the effect of "you know…data! It's…you know…information!" and not venture farther than that. Now it seems to be marketed as the be-all and end-all. The source of not just truth…but intelligence! It's the fuel for artificial intelligence and it is believed that the more data you have, the more truth you have. Therefore, you can never have enough data. It will unlock the secrets needed to redefine your business strategy and maybe even create artificial general intelligence. But let me offer a pragmatic perspective on what data is. Data is not important in itself. It's the analysis of data (and how it is produced) that is the driver of all these innovations and solutions.

Imagine you were provided a photo of a family. Can you glean this family's story based on this one photo? What if you had 20 photos? 200 photos? 2,000 photos? How many photos do you need to know their story? Do you need photos of them in different situations? Alone and together? With relatives and friends? At home and at work?

Data is just like photographs; it provides snapshots of a story. The continuous reality and contexts are not fully captured, nor the infinite number of variables driving that story. As we will discuss, data may be biased. It can have gaps and be missing relevant variables. Ideally, we would love to have an infinite amount of data capturing an infinite number of variables, with so much detail we could virtually re-create reality and construct alternate ones! But is this possible? Currently, no. Not even the greatest supercomputers in the world combined can come close to capturing the entirety of the world as data.

Therefore, we have to narrow our scope to make our objectives feasible. A few strategic photos of the father playing golf can easily tell us whether he is good at golf. But trying to decipher his entire life story just through photos? That might be impossible. There is so much that cannot be captured in snapshots. These practical concerns should also be applied when working with data projects, because data is actually just snapshots of a given time capturing only what it is aimed at (much like a camera). We need to keep our objectives focused as this hones in on gathering data that is relevant and complete. If we make our objectives broad and open-ended, we can get into trouble with spurious findings and incomplete datasets. This practice, known as *data mining*, has a time and place but it must be done carefully. We will revisit this at the end of the chapter.

Even with narrowly defined objectives, we can still run into problems with our data. Let's return to the question of determining whether a few strategic photos can tell whether the father is good at golf. Perhaps if you had a photo of him midswing you would be able to tell if he had good form. Or perhaps if you saw him cheering and fist-pumping at a hole, you can infer he got a good score. Maybe you can just take a photo of his scorecard! But it's important to note all these instances can be faked or taken out of context. Maybe he was cheering for someone else, or maybe the scorecard was not his or even forged. Just like these photographs, data does not capture context or explanations. This is an incredibly important point because data provides clues, not truth. These clues can lead us to the truth or mislead us into erroneous conclusions.

This is why being curious about where data comes from is such an important skill. Ask questions about how the data was created, who created it, and what the data is not capturing. It is too easy to get caught up in what the data says and forget to ask where it came from. Even worse there are widespread sentiments that one can shovel data into machine learning algorithms and expect the computer to work it all out. But as the adage goes, "garbage in, garbage out." No wonder only 13% of machine learning projects succeed, according to VentureBeat (*https://oreil.ly/8hFrO*). Successful machine learning projects put thought and analysis into the data, as well as what produced the data.

Ground Truth

More broadly speaking, the family photo example presents a problem of ground truth (*https://oreil.ly/sa6Ff*).

When I was teaching a class on AI system safety, I was once asked a question on making self-driving cars safer. "When a self-driving car is failing to recognize a pedestrian on its camera sensor, is there not a way for it to recognize the failure and stop?" I replied no, because the system has no framework for *ground truth*, or verified and complete knowledge of what is true. If the car is failing to recognize a pedestrian, how is it supposed to recognize that it is failing to recognize a pedestrian? There's no ground truth for it to fall back on, unless there is a human operator who can provide it and intervene.

This actually is the status quo of "self-driving" cars and taxi services. Some sensors like radar provide modestly reliable ground truths on narrow questions, like "Is there something in front of the vehicle?" But recognizing objects based on camera and LIDAR sensors (in an uncontrolled environment) is a much fuzzier perception problem with astronomical numbers of pixel combinations it could see. Thus, ground truth is nonexistent.

Does your data represent a ground truth that's verifiable and complete? Are the sensors and sources reliable and accurate? Or is the ground truth unknown?

Descriptive Versus Inferential Statistics

What comes to mind when you hear the word "statistics"? Is it calculating mean, median, mode, charts, bell curves, and other tools to describe data? This is the most commonly understood part of statistics, called *descriptive statistics*, and we use it to summarize data. After all, is it more meaningful to scroll through a million records of data or have it summarized? We will cover this area of statistics first.

Inferential statistics tries to uncover attributes about a larger population, often based on a sample. It is often misunderstood and less intuitive than descriptive statistics.

Often we are interested in studying a group that is too large to observe (e.g., average height of adolescents in North America) and we have to resort to using only a few members of that group to infer conclusions about them. As you can guess, this is not easy to get right. After all, we are trying to represent a population with a sample that may not be representative. We will explore these caveats along the way.

Populations, Samples, and Bias

Before we dive deeper into descriptive and inferential statistics, it might be a good idea to lay out some definitions and relate them to tangible examples.

A *population* is a particular group of interest we want to study, such as "all seniors over the age of 65 in the North America," "all golden retrievers in Scotland," or "current high school sophomores at Los Altos High School." Notice how we have boundaries on defining our population. Some of these boundaries are broad and capture a large group over a vast geography or age group. Others are highly specific and small such as the sophomores at Los Altos High School. How you hone in on defining a population depends on what you are interested in studying.

A *sample* is a subset of the population that is ideally random and unbiased, which we use to infer attributes about the population. We often have to study samples because polling the entire population is not always possible. Of course, some populations are easier to get hold of if they are small and accessible. But measuring all seniors over 65 in North America? That is unlikely to be practical!

Populations Can Be Abstract!

It is important to note that populations can be theoretical and not physically tangible. In these cases our population acts more like a sample from something abstract. Here's my favorite example: we are interested in flights that depart between 2 p.m. and 3 p.m. at an airport, but we lack enough flights at that time to reliably predict how often these flights are late. Therefore, we may treat this population as a sample instead from an underlying population of all theoretical flights taking off between 2 p.m. and 3 p.m.

Problems like this are why many researchers resort to simulations to generate data. Simulations can be useful but rarely are accurate, as simulations capture only so many variables and have assumptions built in.

If we are going to infer attributes about a population based on a sample, it's important the sample be as random as possible so we do not skew our conclusions. Here's an example. Let's say I'm a college student at Arizona State University. I want to find the average number of hours college students watch television per week in the United

States. I walk right outside my dorm and start polling random students walking by, finishing my data gathering in a few hours. What's the problem here?

The problem is our student sample is going to have *bias*, meaning it skews our findings by overrepresenting a certain group at the expense of other groups. My study defined the population to be "college students in the United States," not "college students at Arizona State University." I am only polling students at one specific university to represent all college students in the entire United States! Is that really fair?

It is unlikely all colleges across the country homogeneously have the same student attributes. What if Arizona State students watch far more TV than other students at other universities? Would using them to represent the entire country not distort the results? Maybe this is possible because it is usually too hot to go outside in Tempe, Arizona. Therefore TV is a common pastime (anecdotally, I would know; I lived in Phoenix for many years). Other college students in milder climates may do more outdoor activities and watch less TV.

This is just one possible variable showing why it's a bad idea to represent college students across the entire United States with just a sample of students from one university. Ideally, I should be randomly polling college students all over the country at different universities. That way I have a more representative sample.

However, bias is not always geographic. Let's say I make a sincere effort to poll students across the United States. I arrange a social media campaign to have a poll shared by various universities on Twitter and Facebook. This way their students see it and hopefully fill it out. I get hundreds of responses on students' TV habits across the country and feel I've conquered the bias beast...or have I?

What if students who are on social media enough to see the poll are also likely to watch more TV? If they are on social media a lot, they probably do not mind recreational screen time. It's easy to imagine they have Netflix and Hulu ready to stream on that other tab! This particular type of bias where a specific group is more likely to include themselves in a sample is known as *self-selection bias*.

Darn it! You just can't win, can you? If you think about it long enough, data bias just feels inevitable! And often it is. So many *confounding variables*, or factors we did not account for, can influence our study. This problem of data bias is expensive and difficult to overcome, and machine learning is especially vulnerable to it.

The way to overcome this problem is to truly at random select students from the entire population, and they cannot elect themselves into or out of the sample voluntarily. This is the most effective way to mitigate bias, but as you can imagine, it takes a lot of coordinated resources to do.

A Whirlwind Tour of Bias Types

As humans, we are strangely wired to be biased. We look for patterns even if they do not exist. Perhaps this was a necessity for survival in our early history, as finding patterns make hunting, gathering, and farming more productive.

There are many types of bias, but they all have the same effect of distorting findings. *Confirmation bias* is gathering only data that supports your belief, which can even be done unknowingly. An example of this is following only social media accounts you politically agree with, reinforcing your beliefs rather than challenging them.

We just discussed *self-selection bias*, which is when certain types of subjects are more likely to include themselves in the experiment. Walking onto a flight and polling the customers if they like the airline over other airlines, and using that to rank customer satisfaction among all airlines, is silly. Why? Many of those customers are likely repeat customers, and they have created self-selection bias. We do not know how many are first-time fliers or repeat fliers, and of course the latter is going to prefer that airline over other airlines. Even first-time fliers may be biased and self-selected, because they chose to fly that airline and we are not sampling all airlines.

Survival bias captures only living and survived subjects, while the deceased ones are never accounted for. It is the most fascinating type of bias in my opinion, because the examples are diverse and not obvious.

Probably the most famous example of survival bias is this: the Royal Air Force in WWII was having trouble with German enemy fire causing casualties on its bombers. Their initial solution was to armor bombers where bullet holes were found, reasoning this would improve their likelihood of survival. However, a mathematician named Abraham Wald pointed out to them that this was fatally wrong. He proposed armoring areas on the aircraft that *did not* have bullet holes. Was he insane? Far from it. The bombers that returned obviously did not have fatal damage where bullet holes were found. How do we know this? Because they returned from their mission! But what of the aircraft that did not return? Where were they hit? Wald's theory was "likely in the areas untouched in the aircraft *that survived* and returned to base," and this proved to be right. Those areas without bullet holes were armored and the survivability of aircraft and pilots increased. This unorthodox observation is credited with turning the tide of war in favor of the Allies.

There are other fascinating examples of survivor bias that are less obvious. Many management consulting companies and book publishers like to identify traits of successful companies/individuals and use them as predictors for future successes. These works are pure survival bias (and XKCD has a funny cartoon about this (*https://xkcd.com/1827*)). These works do not account for companies/individuals that failed in obscurity, and these "success" qualities may be commonplace with failed ones as well. We just have not heard about them because they never had a spotlight.

An anecdotal example that comes to mind is Steve Jobs. On many accounts, he was said to be passionate and hot-tempered...but he also created one of the most valuable companies of all time. Therefore, some people believed being passionate and even temperamental might correlate with success. Again, this is survivor bias. I am willing to bet there are many companies run by passionate leaders that failed in obscurity, but we fixate and latch on to outlier success stories like Apple.

Lastly, in 1987 there was a veterinary study showing that cats that fell from six or fewer stories had greater injuries than those that fell more than six stories. Prevailing scientific theories postulated that cats righted themselves at about five stories, giving them enough time to brace for impact and inflict less injury. But then The Straight Dope, a column in the *Chicago Reader*, posed an important question: what happened to the dead cats? People are not likely to bring a dead cat to the vet and therefore it is not reported how many cats died from higher falls. In the words of Homer Simpson, "*D'oh!*"

Alright, enough talk about populations, samples, and bias. Let's move on to some math and descriptive statistics. Just remember that math and computers do not recognize bias in your data. That is on you as a good data science professional to detect! Always ask questions about how the data was obtained, and then scrutinize how that process could have biased the data.

Samples and Bias in Machine Learning

These problems with sampling and bias extends to machine learning as well. Whether it is linear regression, logistic regression, or neural networks, a sample of data is used to infer predictions. If that data is biased then it will steer the machine learning algorithm to make biased conclusions.

There are many documented cases of this. Criminal justice has been a precarious application of machine learning because it has repeatedly shown to be biased in every sense of the word, discriminating against minorities due to minority-heavy datasets. In 2017, Volvo tested self-driving cars that were trained on datasets capturing deer, elk, and caribou. However, it had no driving data in Australia and therefore could not recognize kangaroos, much less make sense of their jumping movements! Both of these are examples of biased data.

Descriptive Statistics

Descriptive statistics is the area most people are familiar with. We will touch on the basics like mean, median, and mode followed by variance, standard deviation, and the normal distribution.

Mean and Weighted Mean

The *mean* is the average of a set of values. The operation is simple to do: sum the values and divide by the number of values. The mean is useful because it shows where the "center of gravity" exists for an observed set of values.

The mean is calculated the same way for both populations and samples. Example 3-1 shows a sample of eight values and how to calculate their mean in Python.

Example 3-1. Calculating mean in Python

```
# Number of pets each person owns
sample = [1, 3, 2, 5, 7, 0, 2, 3]

mean = sum(sample) / len(sample)

print(mean) # prints 2.875
```

As you can see, we polled eight people on the number of pets they own. The sum of the sample is 23 and the number of items in the sample is 8, so this gives us a mean of 2.875 as 23/8 = 2.875.

There are two versions of the mean you will see: the sample mean $\bar{x}$ and the population mean μ as expressed here:

$$\bar{x} = \frac{x_1 + x_2 + x_3 + \ldots + x_n}{n} = \sum \frac{x_i}{n}$$

$$\mu = \frac{x_1 + x_2 + x_3 + \ldots + x_n}{N} = \sum \frac{x_i}{N}$$

Recall the summation symbol Σ means add all the items together. The n and the N represent the sample and population size, respectively, but mathematically they represent the same thing: the number of items. The same goes for calling the sample mean $\bar{x}$ ("x-bar") and the population mean μ ("mu"). Both $\bar{x}$ and μ are the same calculation, just different names depending on whether it's a sample or population we are working with.

The mean is likely familiar to you, but here's something less known about the mean: the mean is actually a weighted average called the *weighted mean*. The mean we commonly use gives equal importance to each value. But we can manipulate the mean and give each item a different weight:

$$\text{weighted mean} = \frac{(x_1 \cdot w_1) + (x_2 \cdot w_2) + (x_3 \cdot w_3) + \ldots (x_n \cdot w_n)}{w_1 + w_2 + w_3 + \ldots + w_n}$$

This can be helpful when we want some values to contribute to the mean more than others. A common example of this is weighting academic exams to give a final grade. If you have three exams and a final exam, and we give each of the three exams 20% weight and the final exam 40% weight of the final grade, how we express it is in Example 3-2.

Example 3-2. Calculating a weighted mean in Python

```
# Three exams of .20 weight each and final exam of .40 weight
sample = [90, 80, 63, 87]
weights = [.20, .20, .20, .40]

weighted_mean = sum(s * w for s,w in zip(sample, weights)) / sum(weights)

print(weighted_mean) # prints 81.4
```

We weight each exam score through multiplication accordingly and instead of dividing by the value count, we divide by the sum of weights. Weightings don't have to be percentages, as any numbers used for weights will end up being proportionalized. In Example 3-3, we weight each exam with "1" but weight the final exam with "2," giving it twice the weight of the exams. We will still get the same answer of 81.4 as these values will still be proportionalized.

Example 3-3. Calculating a weighted mean in Python

```
# Three exams of .20 weight each and final exam of .40 weight
sample = [90, 80, 63, 87]
weights = [1.0, 1.0, 1.0, 2.0]

weighted_mean = sum(s * w for s,w in zip(sample, weights)) / sum(weights)

print(weighted_mean) # prints 81.4
```

Median

The *median* is the middlemost value in a set of ordered values. You sequentially order the values, and the median will be the centermost value. If you have an even number of values, you average the two centermost values. We can see in Example 3-4 that the median number of pets owned in our sample is 7:

```
0, 1, 5, *7*, 9, 10, 14
```

Example 3-4. Calculating the median in Python

```python
# Number of pets each person owns
sample = [0, 1, 5, 7, 9, 10, 14]

def median(values):
    ordered = sorted(values)
    print(ordered)
    n = len(ordered)
    mid = int(n / 2) - 1 if n % 2 == 0 else int(n/2)

    if n % 2 == 0:
        return (ordered[mid] + ordered[mid+1]) / 2.0
    else:
        return ordered[mid]

print(median(sample)) # prints 7
```

The median can be a helpful alternative to the mean when data is skewed by *outliers*, or values that are extremely large and small compared to the rest of the values. Here's an interesting anecdote to understand why. In 1986, the mean annual starting salary of geography graduates from the University of North Carolina at Chapel Hill was $250,000. Other universities averaged $22,000. Wow, UNC-CH must have an amazing geography program!

But in reality, what was so lucrative about UNC's geography program? Well…Michael Jordan was one of their graduates. One of the most famous NBA players of all time indeed graduated with a geography degree from UNC. However, he started his career playing basketball, not studying maps. Obviously, this is a confounding variable that has created a huge outlier, and it majorly skewed the income average.

This is why the median can be preferable in outlier-heavy situations (such as income-related data) over the mean. It is less sensitive to outliers and cuts data strictly down the middle based on their relative order, rather than where they fall exactly on a number line. When your median is very different from your mean, that means you have a skewed dataset with outliers.

The Median Is a Quantile

There is a concept of *quantiles* in descriptive statistics. The concept of quantiles is essentially the same as a median, just cutting the data in other places besides the middle. The median is actually the 50% quantile, or the value where 50% of ordered values are behind it. Then there are the 25%, 50%, and 75% quantiles, which are known as *quartiles* because they cut data in 25% increments.

Mode

The *mode* is the most frequently occurring set of values. It primarily becomes useful when your data is repetitive and you want to find which values occur the most frequently.

When no value occurs more than once, there is no mode. When two values occur with an equal amount of frequency, then the dataset is considered *bimodal*. In Example 3-5 we calculate the mode for our pet dataset, and sure enough we see this is bimodal as both 2 and 3 occur the most (and equally) as often.

Example 3-5. Calculating the mode in Python

```
# Number of pets each person owns
from collections import defaultdict

sample = [1, 3, 2, 5, 7, 0, 2, 3]

def mode(values):
    counts = defaultdict(lambda: 0)

    for s in values:
        counts[s] += 1

    max_count = max(counts.values())
    modes = [v for v in set(values) if counts[v] == max_count]
    return modes

print(mode(sample)) # [2, 3]
```

In practicality, the mode is not used a lot unless your data is repetitive. This is commonly encountered with integers, categories, and other discrete variables.

Variance and Standard Deviation

When we start talking about variance and standard deviation, this is where it gets interesting. One thing that confuses people with variance and standard deviation is there are some calculation differences for the sample versus the population. We will do our best to cover these differences clearly.

Population Variance and Standard Deviation

In describing data, we are often interested in measuring the differences between the mean and every data point. This gives us a sense of how "spread out" the data is.

Let's say I'm interested in studying the number of pets owned by members of my work staff (note that I'm defining this as my population, not a sample). I have seven people on my staff.

I take the mean of all the numbers of pets they own, and I get 6.571. Let's subtract this mean from each value. This will show us how far each value is from the mean as shown in Table 3-1.

Table 3-1. Number of pets my staff owns

Value	Mean	Difference
0	6.571	-6.571
1	6.571	-5.571
5	6.571	-1.571
7	6.571	0.429
9	6.571	2.429
10	6.571	3.429
14	6.571	7.429

Let's visualize this on a number line with "X" showing the mean in Figure 3-1.

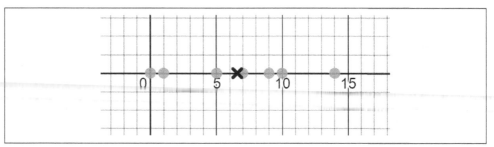

Figure 3-1. Visualizing the spread of our data, where "X" is the mean

Hmmm…now consider why this information can be useful. The differences give us a sense of how spread out the data is and how far values are from the mean. Is there a way we can consolidate these differences into a single number to quickly describe how spread out the data is?

You may be tempted to take the average of the differences, but the negatives and positives will cancel each other out when they are summed. We could sum the absolute values (rid the negative signs and make all values positive). An even better approach would be to square these differences before summing them. This not only rids the negative values (because squaring a negative number makes it positive), but it amplifies larger differences and is mathematically easier to work with (derivatives are not straightforward with absolute values). After that, average the squared differences. This will give us the *variance*, a measure of how spread out our data is.

Here is a math formula showing how to calculate variance:

$$\text{population variance} = \frac{(x_1 - mean)^2 + (x_2 - mean)^2 + \ldots + (x_n - mean)^2}{N}$$

More formally, here is the variance for a population:

$$\sigma^2 = \frac{\Sigma(x_i - \mu)^2}{N}$$

Calculating the population variance of our pet example in Python is shown in Example 3-6.

Example 3-6. Calculating variance in Python

```
# Number of pets each person owns
data = [0, 1, 5, 7, 9, 10, 14]

def variance(values):
    mean = sum(values) / len(values)
    _variance = sum((v - mean) ** 2 for v in values) / len(values)
    return _variance

print(variance(data))  # prints 21.387755102040813
```

So the variance for number of pets owned by my office staff is 21.387755. OK, but what does it exactly mean? It's reasonable to conclude that a higher variance means more spread, but how do we relate this back to our data? This number is larger than any of our observations because we did a lot squaring and summing, putting it on an entirely different metric. So how do we squeeze it back down so it's back on the scale we started with?

The opposite of a square is a square root, so let's take the square root of the variance, which gives us the *standard deviation*. This is the variance scaled into a number expressed in terms of "number of pets," which makes it a bit more meaningful:

$$\sigma = \sqrt{\frac{\Sigma(x_i - \mu)^2}{N}}$$

To implement in Python, we can reuse the `variance()` function and `sqrt()` its result. We now have a `std_dev()` function, shown in Example 3-7.

Example 3-7. Calculating standard deviation in Python

```
from math import sqrt

# Number of pets each person owns
data = [0, 1, 5, 7, 9, 10, 14]

def variance(values):
    mean = sum(values) / len(values)
    _variance = sum((v - mean) ** 2 for v in values) / len(values)
    return _variance

def std_dev(values):
    return sqrt(variance(values))

print(std_dev(data))  # prints 4.624689730353898
```

Running the code in Example 3-7, you will see our standard deviation is approximately 4.62 pets. So we can express our spread on a scale we started with, and this makes our variance a bit easier to interpret. We will see some important applications of the standard deviation in Chapter 5.

Why the Square?

Regarding variance, if the exponent in σ^2 bothers you, it's because it is prompting you to take the square root of it to get the standard deviation. It's a little nagging reminder you are dealing with squared values that need to be square-rooted.

Sample Variance and Standard Deviation

In the previous section we talked about variance and standard deviation for a population. However, there is an important tweak we need to apply to these two formulas when we calculate for a sample:

$$s^2 = \frac{\sum (x_i - \bar{x})^2}{n - 1}$$

$$s = \sqrt{\frac{\sum (x_i - \bar{x})^2}{n - 1}}$$

Did you catch the difference? When we average the squared differences, we divide by $n-1$ rather than the total number of items n. Why would we do this? We do this to decrease any bias in a sample and not underestimate the variance of the population based on our sample. By counting values short of one item in our divisor, we increase the variance and therefore capture greater uncertainty in our sample.

Why Do We Subtract 1 for Sample Size?

On YouTube, Josh Starmer has an excellent video series called StatQuest (*https://oreil.ly/6S9DO*). In one, he gives an excellent breakdown of why we treat samples differently in calculating variance and subtract one element from the number of items.

If our pets data were a sample, not a population, we should make that adjustment accordingly. In Example 3-8, I modify my previous variance() and std_dev() Python code to optionally provide a parameter is_sample, which if True will subtract 1 from the divisor in the variance.

Example 3-8. Calculating standard deviation for a sample

```
from math import sqrt

# Number of pets each person owns
data = [0, 1, 5, 7, 9, 10, 14]

def variance(values, is_sample: bool = False):
    mean = sum(values) / len(values)
    _variance = sum((v - mean) ** 2 for v in values) /
      (len(values) - (1 if is_sample else 0))

    return _variance


def std_dev(values, is_sample: bool = False):
    return sqrt(variance(values, is_sample))

print("VARIANCE = {}".format(variance(data, is_sample=True))) # 24.95238095238095
print("STD DEV = {}".format(std_dev(data, is_sample=True))) # 4.99523582550223
```

Notice in Example 3-8 my variance and standard deviation have increased compared to previous examples that treated them as a population, not a sample. Recall in Example 3-7 that the standard deviation was about 4.62 treating as a population. But here treating as a sample (by subtracting 1 from the variance denominator), we get approximately 4.99. This is correct as a sample could be biased and imperfect

---

representing the population. Therefore, we increase the variance (and thus the standard deviation) to increase our estimate of how spread out the values are. A larger variance/standard deviation shows less confidence with a larger range.

Just like the mean ($\bar{x}$ for sample and μ for population), you will often see certain symbols for variance and standard deviation. The standard deviation for a sample and population are specified by s and σ, respectively. Here again are the sample and population standard deviation formulas:

$$s = \sqrt{\frac{\Sigma (x_i - \bar{x})^2}{n - 1}}$$

$$\sigma = \sqrt{\frac{\Sigma (x_i - \mu)^2}{N}}$$

The variance will be the square of these two formulas, undoing the square root. Therefore, the variance for sample and population are s^2 and σ^2, respectively:

$$s^2 = \frac{\Sigma (x_i - \bar{x})^2}{n - 1}$$

$$\sigma^2 = \frac{\Sigma (x_i - \mu)^2}{N}$$

Again, the square helps imply that a square root should be taken to get the standard deviation.

The Normal Distribution

We touched on probability distributions in the last chapter, particularly the binomial distribution and beta distribution. However the most famous distribution of all is the normal distribution. The *normal distribution*, also known as the *Gaussian distribution*, is a symmetrical bell-shaped distribution that has most mass around the mean, and its spread is defined as a standard deviation. The "tails" on either side become thinner as you move away from the mean.

Figure 3-2 is a normal distribution for golden retriever weights. Notice how most of the mass is around the mean of 64.43 pounds.

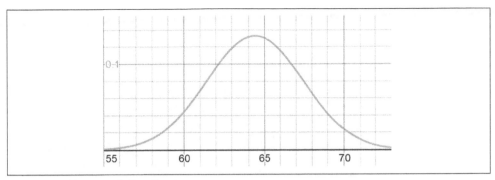

Figure 3-2. A normal distribution

Discovering the Normal Distribution

The normal distribution is seen a lot in nature, engineering, science, and other domains. How do we discover it? Let's say we sample the weight of 50 adult golden retrievers and plot them on a number line as shown in Figure 3-3.

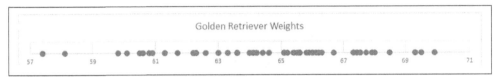

Figure 3-3. A sample of 50 golden retriever weights

Notice how we have more values toward the center, but as we move farther left or right we see fewer values. Based on our sample, it seems highly unlikely we would see a golden retriever with a weight of 57 or 71. But having a weight of 64 or 65? Yeah, that certainly seems likely.

Is there a better way to visualize this likelihood to see which golden retriever weights we are more likely to see sampled from the population? We can try to create a *histogram*, which buckets (or "bins") up values based on numeric ranges of equal length, and then uses a bar chart showing the number of values within each range. In Figure 3-4 we create a histogram that bins up values in ranges of .5 pounds.

This histogram does not reveal any meaningful shape to our data. The reason is because our bins are too small. We do not have an extremely large or infinite amount of data to meaningfully have enough points in each bin. Therefore we will have to make our bins larger. Let's make the bins each have a length of three pounds, as in Figure 3-5.

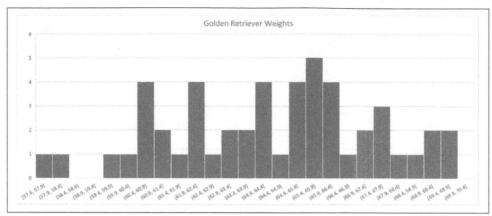

Figure 3-4. A histogram of golden retriever weights

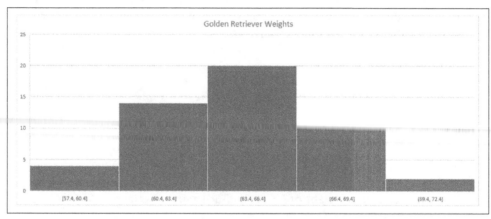

Figure 3-5. A more productive histogram

Now we are getting somewhere! As you can see, if we get the bin sizes just right (in this case, each has a range of three pounds), we start to get a meaningful bell shape to our data. It's not a perfect bell shape because our samples are never going to be perfectly representative of the population, but this is likely evidence our sample follows a normal distribution. If we fit a histogram with adequate bin sizes, and scale it so it has an area of 1.0 (which a probability distribution requires), we see a rough bell curve representing our sample. Let's show it alongside our original data points in Figure 3-6.

Looking at this bell curve, we can reasonably expect a golden retriever to have a weight most likely around 64.43 (the mean) but unlikely at 55 or 73. Anything more extreme than that becomes very unlikely.

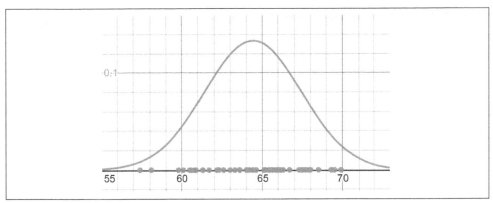

Figure 3-6. A normal distribution fitted to data points

Properties of a Normal Distribution

The normal distribution has several important properties that make it useful:

- It's symmetrical; both sides are identically mirrored at the mean, which is the center.
- Most mass is at the center around the mean.
- It has a spread (being narrow or wide) that is specified by standard deviation.
- The "tails" are the least likely outcomes and approach zero infinitely but never touch zero.
- It resembles a lot of phenomena in nature and daily life, and even generalizes nonnormal problems because of the central limit theorem, which we will talk about shortly.

The Probability Density Function (PDF)

The standard deviation plays an important role in the normal distribution, because it defines how "spread out" it is. It is actually one of the parameters alongside the mean. The *probability density function (PDF)* that creates the normal distribution is as follows:

$$f(x) = \frac{1}{\sigma\sqrt{2\pi}} e^{-\frac{1}{2}\left(\frac{x-\mu}{\sigma}\right)^2}$$

Wow that's a mouthful, isn't it? We even see our friend Euler's Number e from Chapter 1 and some crazy exponents. Here is how we can express it in Python in Example 3-9.

Example 3-9. The normal distribution function in Python

```
# normal distribution, returns likelihood
def normal_pdf(x: float, mean: float, std_dev: float) -> float:
    return (1.0 / (2.0 * math.pi * std_dev ** 2) ** 0.5) *
        math.exp(-1.0 * ((x - mean) ** 2 / (2.0 * std_dev ** 2)))
```

There's a lot to take apart here in this formula, but what's important is that it accepts a mean and standard deviation as parameters, as well as an x-value so you can look up the likelihood at that given value.

Just like the beta distribution in Chapter 2, the normal distribution is continuous. This means to retrieve a probability we need to integrate a range of x values to find an area.

In practice though, we will use SciPy to do these calculations for us.

The Cumulative Distribution Function (CDF)

With the normal distribution, the vertical axis is not the probability but rather the likelihood for the data. To find the probability we need to look at a given range, and then find the area under the curve for that range. Let's say I want to find the probability of a golden retriever weighing between 62 and 66 pounds. Figure 3-7 shows the range we want to find the area for.

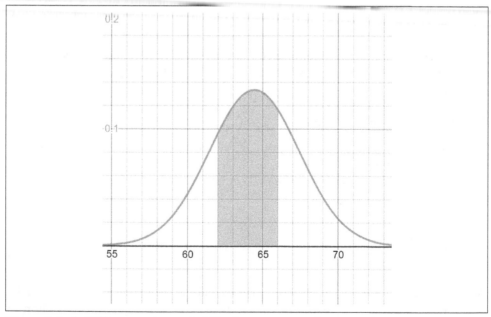

Figure 3-7. A CDF measuring probability between 62 and 66 pounds

We already did this task in Chapter 2 with the beta distribution, and just like the beta distribution there is a cumulative density function (CDF). Let's follow this approach.

As we learned in the last chapter, the CDF provides the area *up to* a given x-value for a given distribution. Let's see what the CDF looks like for our golden retriever normal distribution and put it alongside the PDF for reference in Figure 3-8.

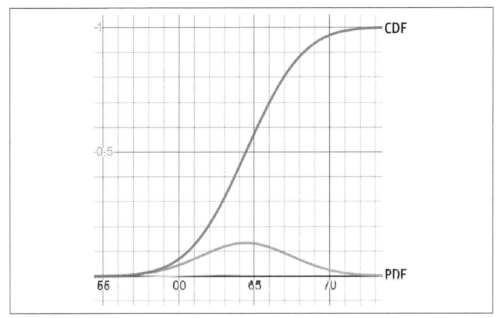

Figure 3-8. A PDF alongside its CDF

Notice there's a relationship between the two graphs. The CDF, which is an S-shaped curve (called a sigmoid curve), projects the area up to that range in the PDF. Observe in Figure 3-9 that when we capture the area from negative infinity up to 64.43 (the mean), our CDF shows a value of exactly .5 or 50%!

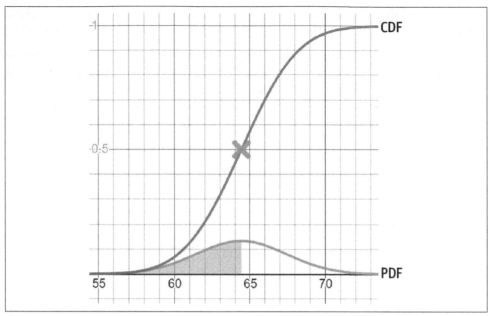

Figure 3-9. A PDF and CDF for golden retriever weights measuring probability up to the mean

This area of .5 or 50% up to the mean is known because of the symmetry of our normal distribution, and we can expect the other side of the bell curve to also have 50% of the area.

To calculate this area up to 64.43 in Python using SciPy, use the `norm.cdf()` function as shown in Example 3-10.

Example 3-10. The normal distribution CDF in Python

```
from scipy.stats import norm

mean = 64.43
std_dev = 2.99

x = norm.cdf(64.43, mean, std_dev)

print(x) # prints 0.5
```

Just like we did in Chapter 2, we can deductively find the area for a middle range by subtracting areas. If we wanted to find the probability of observing a golden retriever between 62 and 66 pounds, we would calculate the area up to 66 and subtract the area up to 62 as visualized in Figure 3-10.

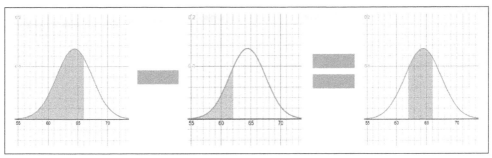

Figure 3-10. Finding a middle range of probability

Doing this in Python using SciPy is as simple as subtracting the two CDF operations shown in Example 3-11.

Example 3-11. Getting a middle range probability using the CDF

```
from scipy.stats import norm

mean = 64.43
std_dev = 2.99

x = norm.cdf(66, mean, std_dev) - norm.cdf(62, mean, std_dev)

print(x) # prints 0.4920450147062894
```

You should find the probability of observing a golden retriever between 62 and 66 pounds to be 0.4920, or approximately 49.2%.

The Inverse CDF

When we start doing hypothesis testing later in this chapter, we will encounter situations where we need to look up an area on the CDF and then return the corresponding x-value. Of course this is a backward usage of the CDF, so we will need to use the inverse CDF, which flips the axes as shown in Figure 3-11.

This way, we can now look up a probability and then return the corresponding x-value, and in SciPy we would use the norm.ppf() function. For example, I want to find the weight that 95% of golden retrievers fall under. This is easy to do when I use the inverse CDF in Example 3-12.

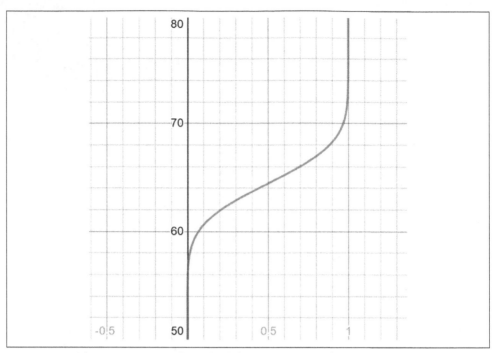

Figure 3-11. The inverse CDF, also called the PPF or quantile function

Example 3-12. Using the inverse CDF (called ppf()) in Python

```
from scipy.stats import norm

x = norm.ppf(.95, loc=64.43, scale=2.99)
print(x) # 69.3481123445849
```

I find that 95% of golden retrievers are 69.348 or fewer pounds.

You can also use the inverse CDF to generate random numbers that follow the normal distribution. If I want to create a simulation that generates one thousand realistic golden retriever weights, I just generate a random value between 0.0 and 1.0, pass it to the inverse CDF, and return the weight value as shown in Example 3-13.

Example 3-13. Generating random numbers from a normal distribution

```
import random
from scipy.stats import norm

for i in range(0,1000):
    random_p = random.uniform(0.0, 1.0)
    random_weight = norm.ppf(random_p,  loc=64.43, scale=2.99)
    print(random_weight)
```

Of course NumPy and other libraries can generate random values off a distribution for you, but this highlights one use case where the inverse CDF is handy.

CDF and Inverse CDF from Scratch

To learn how to implement the CDF and inverse CDF from scratch in Python, refer to Appendix A.

Z-Scores

It is common to rescale a normal distribution so that the mean is 0 and the standard deviation is 1, which is known as the *standard normal distribution*. This makes it easy to compare the spread of one normal distribution to another normal distribution, even if they have different means and variances.

Of particular importance with the standard normal distribution is it expresses all x-values in terms of standard deviations, known as *Z-scores*. Turning an x-value into a Z-score uses a basic scaling formula:

$$z = \frac{x - \mu}{\sigma}$$

Here is an example. We have two homes from two different neighborhoods. Neighborhood A has a mean home value of \$140,000 and standard deviation of \$3,000. Neighborhood B has a mean home value of \$800,000 and standard deviation of \$10,000.

$\mu_A = 140{,}000$
$\mu_B = 800{,}000$
$\sigma_A = 3{,}000$
$\sigma_B = 10{,}000$

Now we have two homes from each neighborhood. House A from neighborhood A is worth \$150,000 and house B from neighborhood B is worth \$815,000. Which home is more expensive relative to the average home in its neighborhood?

$x_A = 150{,}000$
$x_B = 815{,}000$

If we express these two values in terms of standard deviations, we can compare them relative to each neighborhood mean. Use the Z-score formula:

$$z = \frac{x - \text{mean}}{\text{standard deviation}}$$

$$z_A = \frac{150000 - 140000}{3000} = 3.\overline{333}$$

$$z_B = \frac{815000 - 800000}{10000} = 1.5$$

So the house in neighborhood A is actually much more expensive relative to its neighborhood than the house in neighborhood B, as they have Z-scores of $3.\overline{333}$ and 1.5, respectively.

Here is how we can convert an x-value coming from a given distribution with a mean and standard deviation into a Z-score, and vice versa, as shown in Example 3-14.

Example 3-14. Turn Z-scores into x-values and vice versa

```
def z_score(x, mean, std):
    return (x - mean) / std


def z_to_x(z, mean, std):
    return (z * std) + mean


mean = 140000
std_dev = 3000
x = 150000

# Convert to Z-score and then back to X
z = z_score(x, mean, std_dev)
back_to_x = z_to_x(z, mean, std_dev)

print("Z-Score: {}".format(z))  # Z-Score: 3.333
print("Back to X: {}".format(back_to_x))  # Back to X: 150000.0
```

The `z_score()` function will take an x-value and scale it in terms of standard deviations, given a mean and standard deviation. The `z_to_x()` function takes a Z-score and converts it back to an x-value. Studying the two functions, you can see their algebraic relationship, one solving for the Z-score and the other for the x-value. We then turn an x-value of 8.0 into a Z-score of $3.\overline{333}$ and then turn that Z-score back into an x-value.

Coefficient of Variation

A helpful tool for measuring spread is the coefficient of variation. It compares two distributions and quantifies how spread out each of them is. It is simple to calculate: divide the standard deviation by the mean. Here is the formula alongside the example comparing two neighborhoods:

$$cv = \frac{\sigma}{\mu}$$

$$cv_A = \frac{3000}{140000} = 0.0214$$

$$cv_B = \frac{10000}{800000} = 0.0125$$

As seen here, neighborhood A, while cheaper than neighborhood B, has more spread and therefore more price diversity than neighborhood B.

Inferential Statistics

Descriptive statistics, which we have covered so far, is commonly understood. However, when we get into inferential statistics the abstract relationships between sample and population come into full play. These abstract nuances are not something you want to rush through but rather take your time and absorb thoughtfully. As stated earlier, we are wired as humans to be biased and quickly come to conclusions. Being a good data science professional requires you to suppress that primal desire and consider the possibility that other explanations can exist. It is acceptable (perhaps even enlightened) to theorize there is no explanation at all and a finding is just coincidental and random.

First let's start with the theorem that lays the foundation for all inferential statistics.

The Central Limit Theorem

One of the reasons the normal distribution is useful is because it appears a lot in nature, such as adult golden retriever weights. However, it shows up in a more fascinating context outside of natural populations. When we start measuring large enough samples from a population, even if that population does not follow a normal distribution, the normal distribution still makes an appearance.

Let's pretend I am measuring a population that is truly and uniformly random. Any value between 0.0 and 1.0 is equally likely, and no value has any preference. But something fascinating happens when we take increasingly large samples from this population, take the average of each, and then plot them into a histogram. Run this Python code in Example 3-15 and observe the plot in Figure 3-12.

Example 3-15. Exploring the central limit theorem in Python

```
# Samples of the uniform distribution will average out to a normal distribution.
import random
import plotly.express as px

sample_size = 31
sample_count = 1000

# Central limit theorem, 1000 samples each with 31
# random numbers between 0.0 and 1.0
x_values = [(sum([random.uniform(0.0, 1.0) for i in range(sample_size)])) / \
    sample_size)
            for _ in range(sample_count)]

y_values = [1 for _ in range(sample_count)]

px.histogram(x=x_values, y = y_values, nbins=20).show()
```

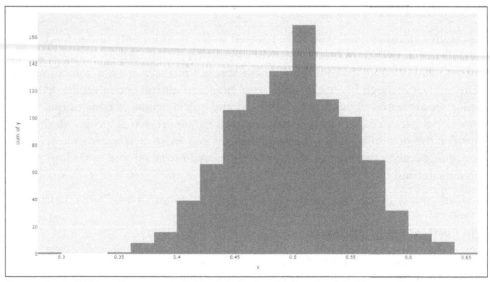

Figure 3-12. Taking the means of samples (each of size 31) and plotting them

Wait, how did uniformly random numbers, when sampled as groups of 31 and then averaged, roughly form a normal distribution? Any number is equally likely, right? Shouldn't the distribution be flat rather than bell-curved?

Here's what is happening. The individual numbers in the samples alone will not create a normal distribution. The distribution will be flat where any number is equally likely (known as a *uniform distribution*). But when we group them as samples and average them, they form a normal distribution.

This is because of the *central limit theorem*, which states that interesting things happen when we take large enough samples of a population, calculate the mean of each, and plot them as a distribution:

1. The mean of the sample means is equal to the population mean.
2. If the population is normal, then the sample means will be normal.
3. If the population is not normal, but the sample size is greater than 30, the sample means will still roughly form a normal distribution.
4. The standard deviation of the sample means equals the population standard deviation divided by the square root of *n*:

$$\sigma_{\bar{x}} = \frac{\sigma}{\sqrt{n}}$$

Why is all of the above important? These behaviors allows us to infer useful things about populations based on samples, even for nonnormal populations. If you modify the preceding code and try smaller sample sizes of 1 or 2, you will not see a normal distribution emerge. But as you approach 31 or more, you will see that we converge onto a normal distribution as shown in Figure 3-13.

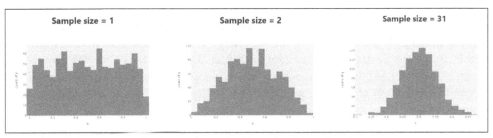

Figure 3-13. Larger sample sizes approach the normal distribution

Thirty-one is the textbook number in statistics because that is when our sample distribution often converges onto the population distribution, particularly when we measure sample means or other parameters. When you have fewer than 31 items in your sample, that is when you have to rely on the T-distribution rather than the normal distribution, which has increasingly fatter tails the smaller your sample size. We will briefly talk about this later, but first let's assume we have at least 31 items in our samples when we talk about confidence intervals and testing.

How Much Sample Is Enough?

While 31 is the textbook number of items you need in a sample to satisfy the central limit theorem and see a normal distribution, this sometimes is not the case. There are cases when you will need an even larger sample, such as when the underlying distribution is asymmetrical or multimodal (meaning it has several peaks rather than one at the mean).

In summary, having larger samples is better when you are uncertain of the underlying probability distribution. You can read more in this article (*https://oreil.ly/IZ4Rk*).

Confidence Intervals

You may have heard the term "confidence interval," which often confuses statistics newcomers and students. A *confidence interval* is a range calculation showing how confidently we believe a sample mean (or other parameter) falls in a range for the population mean.

Based on a sample of 31 golden retrievers with a sample mean of 64.408 and a sample standard deviation of 2.05, I am 95% confident that the population mean lies between 63.686 and 65.1296. How do I know this? Let me show you, and if you get confused, circle back to this paragraph and remember what we are trying to achieve. I highlighted it for a reason!

I first start out by choosing a *level of confidence (LOC)*, which will contain the desired probability for the population mean range. I want to be 95% confident that my sample mean falls in the population mean range I will calculate. That's my LOC. We can leverage the central limit theorem and infer what this range for the population mean is. First, I need the *critical z-value* which is the symmetrical range in a standard normal distribution that gives me 95% probability in the center as highlighted in Figure 3-14.

How do we calculate this symmetrical range containing .95 of the area? It's easier to grasp as a concept than as a calculation. You may instinctively want to use the CDF, but then you may realize there are a few more moving parts here.

First you need to leverage the inverse CDF. Logically, to get 95% of the symmetrical area in the center, we would chop off the tails that have the remaining 5% of area. Splitting that remaining 5% area in half would give us 2.5% area in each tail. Therefore, the areas we want to look up the x-values for are .025 and .975 as shown in Figure 3-15.

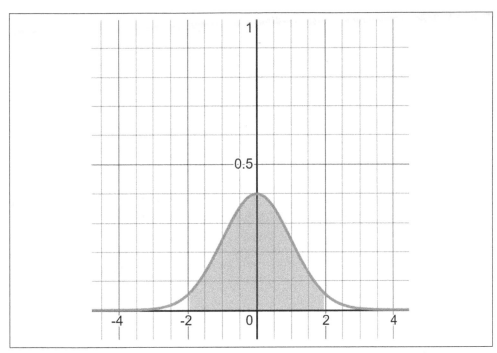

Figure 3-14. 95% symmetrical probability in the center of a standard normal distribution

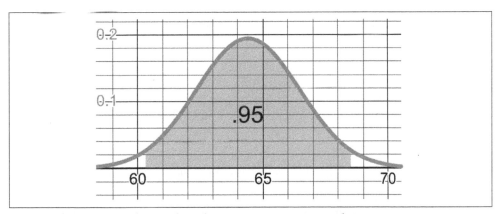

Figure 3-15. We want the x-values that give us areas .025 and .975

We can look up the x-value for area .025 and the x-value for area .975, and that will give us our center range containing 95% of the area. We will then return the corresponding lower and upper z-values containing this area. Remember, we are using standard normal distribution here so they will be the same other than being positive/negative. Let's calculate this in Python as shown in Example 3-16.

Example 3-16. Retrieving a critical z-value

```
from scipy.stats import norm

def critical_z_value(p):
    norm_dist = norm(loc=0.0, scale=1.0)
    left_tail_area = (1.0 - p) / 2.0
    upper_area = 1.0 - ((1.0 - p) / 2.0)
    return norm_dist.ppf(left_tail_area), norm_dist.ppf(upper_area)

print(critical_z_value(p=.95))
# (-1.959963984540054, 1.959963984540054)
```

OK, so we get ±1.95996, which is our critical z-value capturing 95% of probability at the center of the standard normal distribution. Next I'm going to leverage the central limit theorem to produce the *margin of error (E)*, which is the range around the sample mean that contains the population mean at that level of confidence. Recall that our sample of 31 golden retrievers has a mean of 64.408 and standard deviation of 2.05. The formula to get this margin of error is:

$$E = \pm z_c \frac{s}{\sqrt{n}}$$
$$E = \pm 1.95996 * \frac{2.05}{\sqrt{31}}$$
$$E = \pm 0.72164$$

If we apply that margin of error against the sample mean, we finally get the confidence interval!

95% confidence interval = 64.408 ± 0.72164

Here is how we calculate this confidence interval in Python from beginning to end in Example 3-17.

Example 3-17. Calculating a confidence interval in Python

```
from math import sqrt
from scipy.stats import norm


def critical_z_value(p):
    norm_dist = norm(loc=0.0, scale=1.0)
    left_tail_area = (1.0 - p) / 2.0
    upper_area = 1.0 - ((1.0 - p) / 2.0)
    return norm_dist.ppf(left_tail_area), norm_dist.ppf(upper_area)
```

```
def confidence_interval(p, sample_mean, sample_std, n):
    # Sample size must be greater than 30

    lower, upper = critical_z_value(p)
    lower_ci = lower * (sample_std / sqrt(n))
    upper_ci = upper * (sample_std / sqrt(n))

    return sample_mean - lower_ci, sample_mean + upper_ci

print(confidence_interval(p=.95, sample_mean=64.408, sample_std=2.05, n=31))
# (63.68635915701992, 65.12964084298008)
```

So the way to interpret this is "based on my sample of 31 golden retriever weights with sample mean 64.408 and sample standard deviation of 2.05, I am 95% confident the population mean lies between 63.686 and 65.1296." That is how we describe our confidence interval.

One interesting thing to note here too is that in our margin of error formula, the larger n becomes, the narrower our confidence interval becomes! This makes sense because if we have a larger sample, we are more confident in the population mean falling in a smaller range, hence why it's called a confidence interval.

One caveat to put here is that for this to work, our sample size must be at least 31 items. This goes back to the central limit theorem. If we want to apply a confidence interval to a smaller sample, we need to use a distribution with higher variance (fatter tails reflecting more uncertainty). This is what the T-distribution is for, and we will visit that at the end of this chapter.

In Chapter 5 we will continue to use confidence intervals for linear regressions.

Understanding P-Values

When we say something is *statistically significant*, what do we mean by that? We hear it used loosely and frequently but what does it mean mathematically? Technically, it has to do with something called the p-value, which is a hard concept for many folks to grasp. But I think the concept of p-values makes more sense when you trace it back to its invention. While this is an imperfect example, it gets across some big ideas.

In 1925, mathematician Ronald Fisher was at a party. One of his colleagues Muriel Bristol claimed she could detect when tea was poured before milk simply by tasting it. Intrigued by the claim, Ronald set up an experiment on the spot.

He prepared eight cups of tea. Four had milk poured first; the other four had tea poured first. He then presented them to his connoisseur colleague and asked her to identify the pour order for each. Remarkably, she identified them all correctly, and the probability of this happening by chance is 1 in 70, or 0.01428571.

This 1.4% probability is what we call the *p-value*, the probability of something occurring by chance rather than because of a hypothesized explanation. Without going down a rabbit hole of combinatorial math, the probability that Muriel completely guessed the cups correctly is 1.4%. What exactly does that tell you?

When we frame an experiment, whether it is determining if organic donuts cause weight gain or living near power lines causes cancer, we always have to entertain the possibility that random luck played a role. Just like there is a 1.4% chance Muriel identified the cups of tea correctly simply by guessing, there's always a chance randomness just gave us a good hand like a slot machine. This helps us frame our *null hypothesis* (H_0), saying that the variable in question had no impact on the experiment and any positive results are just random luck. The *alternative hypothesis* (H_1) poses that a variable in question (called the *controlled variable*) is causing a positive result.

Traditionally, the threshold for statistical significance is a p-value of 5% or less, or .05. Since .014 is less than .05, this would mean we can reject our null hypothesis that Muriel was randomly guessing. We can then promote the alternative hypothesis that Muriel had a special ability to detect whether tea or milk was poured first.

Now one thing this tea-party example did not capture is that when we calculate a p-value, we capture all probability of that event or rarer. We will address this as we dive into the next example using the normal distribution.

Hypothesis Testing

The population of people with a cold has a mean recovery time of 18 days, with a standard deviation of 1.5 days, and follows a normal distribution.

This means there is approximately 95% chance of recovery taking between 15 and 21 days as shown in Figure 3-16 and Example 3-18.

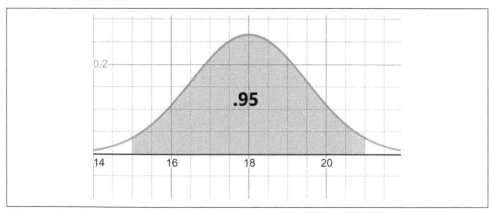

Figure 3-16. There is 95% chance of recovery between 15 and 21 days

Example 3-18. Calculating the probability of recovery between 15 and 21 days

```
from scipy.stats import norm

# Cold has 18 day mean recovery, 1.5 std dev
mean = 18
std_dev = 1.5

# 95% probability recovery time takes between 15 and 21 days.
x = norm.cdf(21, mean, std_dev) - norm.cdf(15, mean, std_dev)

print(x) # 0.9544997361036416
```

We can infer then from the remaining 5% probability that there's a 2.5% chance of recovery taking longer than 21 days and a 2.5% chance of it taking fewer than 15 days. Hold onto that bit of information because it will be critical later! That drives our p-value.

Now let's say an experimental new drug was given to a test group of 40 people, and it took an average of 16 days for them to recover from the cold as shown in Figure 3-17.

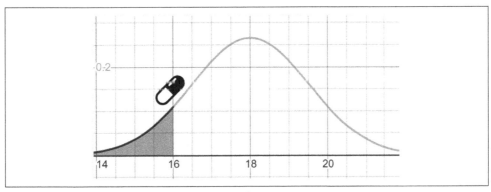

Figure 3-17. A group taking a drug took 16 days to recover

Did the drug have an impact? If you reason long enough, you may realize what we are asking is this: does the drug show a statistically signficant result? Or did the drug not work and the 16-day recovery was a coincidence with the test group? That first question frames our alternative hypothesis, while the second question frames our null hypothesis.

There are two ways we can calculate this: the one-tailed and two-tailed test. We will start with the one-tailed.

One-Tailed Test

When we approach the *one-tailed test*, we typically frame our null and alternative hypotheses using inequalities. We hypothesize around the population mean and say that it either is greater than/equal to 18 (the null hypothesis H_0) or less than 18 (the alternative hypothesis H_1):

H_0:population mean $\geq$ 18
H_1:population mean $<$ 18

To reject our null hypothesis, we need to show that our sample mean of the patients who took the drug is not likely to have been coincidental. Since a p-value of .05 or less is traditionally considered statistically signficant, we will use that as our threshold (Figure 3-17). When we calculate this in Python using the inverse CDF as shown in Figure 3-18 and Example 3-19, we find that approximately 15.53 is the number of recovery days that gives us .05 area on the left tail.

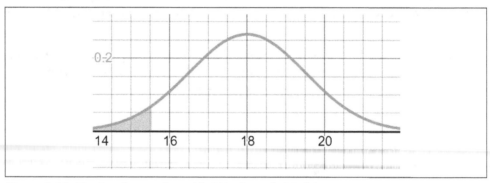

Figure 3-18. Getting the x-value with 5% of area behind it

Example 3-19. Python code for getting x-value with 5% of area behind it

```
from scipy.stats import norm

# Cold has 18 day mean recovery, 1.5 std dev
mean = 18
std_dev = 1.5

# What x-value has 5% of area behind it?
x = norm.ppf(.05, mean, std_dev)

print(x) # 15.53271955957279
```

Therefore, if we achieve an average 15.53 or fewer days of recovery time in our sample group, our drug is considered statistically significant enough to have shown an impact. However, our sample mean of recovery time is actually 16 days and does not fall into this null hypothesis rejection zone. Therefore, the statistical significance test has failed as shown in Figure 3-19.

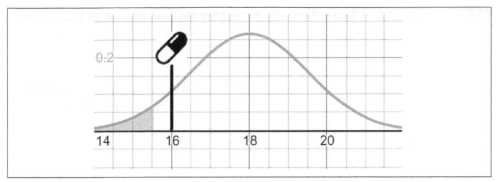

Figure 3-19. We have failed to prove our drug test result is statistically significant

The area up to that 16-day mark is our p-value, which is .0912, and we calculate it in Python as shown in Example 3-20.

Example 3-20. Calculating the one-tailed p-value

```
from scipy.stats import norm

# Cold has 18 day mean recovery, 1.5 std dev
mean = 18
std_dev = 1.5

# Probability of 16 or less days
p_value = norm.cdf(16, mean, std_dev)

print(p_value) # 0.09121121972586788
```

Since the p-value of .0912 is greater than our statistical significance threshold of .05, we do not consider the drug trial a success and fail to reject our null hypothesis.

Two-Tailed Test

The previous test we performed is called the one-tailed test because it looks for statistical significance only on one tail. However, it is often safer and better practice to use a two-tailed test. We will elaborate why, but first let's calculate it.

To do a *two-tailed test*, we frame our null and alternative hypothesis in an "equal" and "not equal" structure. In our drug test, we will say the null hypothesis has a mean recovery time of 18 days. But our alternative hypothesis is the mean recovery time is not 18 days, thanks to the new drug:

H_0: population mean = 18
H_1: population mean ≠ 18

This has an important implication. We are structuring our alternative hypothesis to not test whether the drug improves cold recovery time, but if it had *any* impact. This includes testing if it increased the duration of the cold. Is this helpful? Hold that thought.

Naturally, this means we spread our p-value statistical significance threshold to both tails, not just one. If we are testing for a statistical significance of 5%, then we split it and give each 2.5% half to each tail. If our drug's mean recovery time falls in either region, our test is successful and we reject the null hypothesis (Figure 3-20).

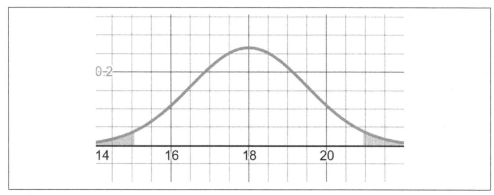

Figure 3-20. A two-tailed test

The x-values for the lower tail and upper tail are 15.06 and 20.93, meaning if we are under or over, respectively, we reject the null hypothesis. Those two values are calculated using the inverse CDF shown in Figure 3-21 and Example 3-21. Remember, to get the upper tail we take .95 and then add the .025 piece of significance threshold to it, giving us .975.

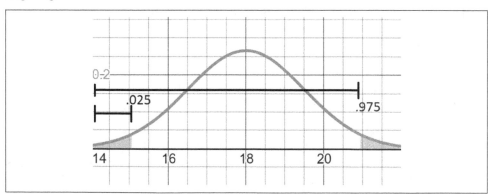

Figure 3-21. Calculating the central 95% of normal distribution area

Example 3-21. Calculating a range for a statistical significance of 5%

```
from scipy.stats import norm

# Cold has 18 day mean recovery, 1.5 std dev
mean = 18
std_dev = 1.5

# What x-value has 2.5% of area behind it?
x1 = norm.ppf(.025, mean, std_dev)

# What x-value has 97.5% of area behind it
x2 = norm.ppf(.975, mean, std_dev)

print(x1) # 15.060054023189918
print(x2) # 20.93994597681008
```

The sample mean value for the drug test group is 16, and 16 is not less than 15.06 nor greater than 20.9399. So like the one-tailed test, we still fail to reject the null hypothesis. Our drug still has not shown any statistical significance to have any impact as shown in Figure 3-22.

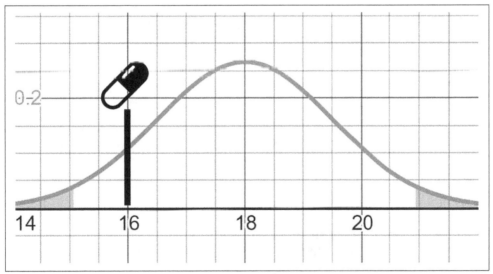

Figure 3-22. The two-tailed test has failed to prove statistical significance

But what is the p-value? This is where it gets interesting with two-tailed tests. Our p-value is going to capture not just the area to the left of 16 but also the symmetrical equivalent area on the right tail. Since 16 is 2 days below the mean, we will also capture the area above 20, which is 2 days above the mean (Figure 3-23). This is capturing the probability of an event or rarer, on both sides of the bell curve.

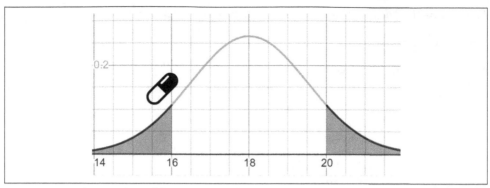

Figure 3-23. The p-value adds symmetrical sides for statistical significance

When we sum both those areas, we get a p-value of .1824. This is a lot greater than .05, so it definitely does not pass our p-value threshold of .05 (Example 3-22).

Example 3-22. Calculating the two-tailed p-value

```
from scipy.stats import norm

# Cold has 18 day mean recovery, 1.5 std dev
mean = 18
std_dev = 1.5

# Probability of 16 or less days
p1 = norm.cdf(16, mean, std_dev)

# Probability of 20 or more days
p2 = 1.0 -  norm.cdf(20, mean, std_dev)

# P-value of both tails
p_value = p1 + p2

print(p_value) # 0.18242243945173575
```

So why do we also add the symmetrical area on the opposite side in a two-tailed test? This may not be the most intuitive concept, but first remember how we structured our hypotheses:

H_0:population mean = 18
H_1:population mean ≠ 18

If we are testing in an "equals 18" versus "not equals 18" capacity, we have to capture any probability that is of equal or less value on both sides. After all, we are trying to prove significance, and that includes anything that is equally or less likely to happen. We did not have this special consideration with the one-tailed test that used only "greater/less than" logic. But when we are dealing with "equals/not equals" our interest area goes in both directions.

So what are the practical implications of the two-tailed test? How does it affect whether we reject the null hypothesis? Ask yourself this: which one sets a higher threshold? You will notice that even when our objective is to show we may have lessened something (the cold-recovery time using a drug), reframing our hypothesis to show any impact (greater or lesser) creates a higher significance threshold. If our significance threshold is a p-value of .05 or less, our one-tailed test was closer to acceptance at p-value .0912 as opposed to the two-tailed test, which was about double that at p-value .182.

This means the two-tailed test makes it harder to reject the null hypothesis and demands stronger evidence to pass a test. Also think of this: what if our drug could worsen colds and make them last longer? It may be helpful to capture that probability too and account for variation in that direction. This is why two-tailed tests are preferable in most cases. They tend to be more reliable and not bias the hypothesis in just one direction.

We will use hypothesis testing and p-values again in Chapters 5 and 6.

Beware of P-Hacking!

There is a problem getting more awareness in the scientific research community called p-hacking, where researchers shop for statistically significant p-values of .05 or less. This is not hard to do with big data, machine learning, and data mining where we can traverse hundreds or thousands of variables and then find statistically significant (but coincidental) relationships between them.

Why do so many researchers p-hack? Many are probably not aware they are doing it. It's easy to keep tuning and tuning a model, omitting "noisy" data and changing parameters, until it gives the "right" result. Others are simply under pressure by academia and industry to produce lucrative, not objective, results.

If you are hired by the Calvin Cereal Company to study whether Frosted Chocolate Sugar Bombs causes diabetes, do you expect to give an honest analysis and get hired again? What if a manager asks you to produce a forecast showing $15 million in sales next quarter for a new product launch? You have no control over what sales will be, but you're asked for a model that produces a predetermined result. In the worst case, you may even be held accountable when it proves wrong. Unfair, but it happens!

This is why soft skills like diplomacy can make a difference in a data science professional's career. If you can share difficult and inconvenient stories in a way that productively gets an audience; that is a tremendous feat. Just be mindful of the organization's managerial climate and always propose alternate solutions. If you end up in a no-win situation where you are asked to p-hack, and could be held accountable for when it backfires, then definitely change work environments!

The T-Distribution: Dealing with Small Samples

Let's briefly address how to deal with smaller samples of 30 or fewer; we will need this when we do linear regression in Chapter 5. Whether we are calculating confidence intervals or doing hypothesis testing, if we have 30 or fewer items in a sample we would opt to use a T-distribution instead of a normal distribution. The *T-distribution* is like a normal distribution but has fatter tails to reflect more variance and uncertainty. Figure 3-24 shows a normal distribution (dashed) alongside a T-distribution with one degree of freedom (solid).

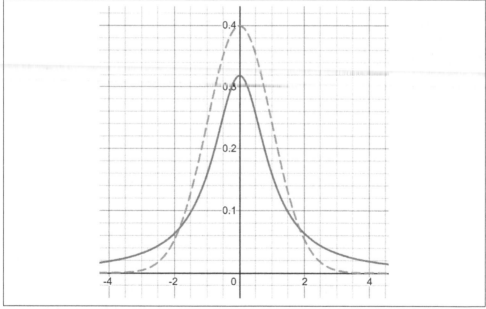

Figure 3-24. The T-distribution alongside a normal distribution; note the fatter tails

The smaller the sample size, the fatter the tails get in a T-distribution. But what's interesting is after you approach 31 items, the T-distribution is nearly indistinguishable from the normal distribution, which neatly reflects the ideas behind the central limit theorem.

Example 3-23 shows how to find the *critical t-value* for 95% confidence. You can use this for confidence intervals and hypothesis testing when you have a sample size of 30 or less. It's conceptually the same as the critical z-value, but we are using a T-distribution instead of a normal distribution to reflect greater uncertainty. The smaller the sample size, the larger the range, reflecting greater uncertainty.

Example 3-23. Getting a critical value range with a T-distribution

```
from scipy.stats import t

# get critical value range for 95% confidence
# with a sample size of 25

n = 25
lower = t.ppf(.025, df=n-1)
upper = t.ppf(.975, df=n-1)

print(lower, upper)
-2.063898561628021 2.0638985616280205
```

Note that df is the "degrees of freedom" parameter, and as outlined earlier it should be one less of the sample size.

Beyond the Mean

We can use confidence intervals and hypothesis testing to measure other parameters besides the mean, including variance/standard deviation as well as proportions (e.g., 60% of people report improved happiness after walking one hour a day). These require other distributions such as the chi-square distribution instead of the normal distribution. As stated, these are all beyond the scope of this book but hopefully you have a strong conceptual foundation to extend into these areas if needed.

We will, however, use confidence intervals and hypothesis testing throughout Chapters 5 and 6.

Big Data Considerations and the Texas Sharpshooter Fallacy

One final thought before we close this chapter. As we have discussed, randomness plays such a role in validating our findings and we always have to account for its possibility. Unfortunately with big data, machine learning, and other data-mining tools, the scientific method has suddenly become a practice done backward. This can be precarious; allow me to demonstrate why, adapting an example from Gary Smith's book *Standard Deviations* (Overlook Press).

Let's pretend I draw four playing cards from a fair deck. There's no game or objective here other than to draw four cards and observe them. I get two 10s, a 3, and a 2. "This is interesting," I say. "I got two 10s, a 3, and a 2. Is this meaningful? Are the next four cards I draw also going to be two consecutive numbers and a pair? What's the underlying model here?"

See what I did there? I took something that was completely random and I not only looked for patterns, but I tried to make a predictive model out of them. What has subtly happened here is I never made it my objective to get these four cards with these particular patterns. I observed them *after* they occurred.

This is exactly what data mining falls victim to every day: finding coincidental patterns in random events. With huge amounts of data and fast algorithms looking for patterns, it's easy to find things that look meaningful but actually are just random coincidences.

This is also analogous to me firing a gun at a wall. I then draw a target around the hole and bring my friends over to show off my amazing marksmanship. Silly, right? Well, many people in data science figuratively do this every day and it is known as the *Texas Sharpshooter Fallacy*. They set out to act without an objective, stumble on something rare, and then point out that what they found somehow creates predictive value.

The problem is the law of truly large numbers says rare events are likely to be found; we just do not know which ones. When we encounter rare events, we highlight and even speculate what might have caused them. The issue is this: the probability of a specific person winning the lottery is highly unlikely, but yet someone *is* going to win the lottery. Why should we be surprised when there is a winner? Unless somebody predicted the winner, nothing meaningful happened other than a random person got lucky.

This also applies to correlations, which we will study in Chapter 5. With an enormous dataset with thousands of variables, is it easy to find statistically significant findings with a .05 p-value? You betcha! I'll find thousands of those. I'll even show evidence that the number of Nicolas Cage movies correlates with the number of pool drownings in a year (*https://oreil.ly/eGxm0*).

So to prevent the Texas Sharpshooter Fallacy and falling victim to big data fallacies, try to use structured hypothesis testing and gather data for that objective. If you utilize data mining, try to obtain fresh data to see if your findings still hold up. Finally, always consider the possibility that things can be coincidental; if there is not a commonsense explanation, then it probably was coincidental.

We learned how to hypothesize before gathering data, but data mining gathers data, then hypothesizes. Ironically, we are often more objective starting with a hypothesis, because we then seek out data to deliberately prove and disprove our hypothesis.

Conclusion

We learned a lot in this chapter, and you should feel good about getting this far. This was probably one of the harder topics in this book! We not only learned descriptive statistics from the mean to the normal distribution but also tackled confidence intervals and hypothesis testing.

Hopefully, you see data a little differently. It is snapshots of something rather than a complete capturing of reality. Data on its own is not very useful, and we need context, curiosity, and analysis of where it came from before we can make meaningful insights with it. We covered how to describe data as well as infer attributes about a larger population based on a sample. Finally, we addressed some of the data-mining fallacies we can stumble into if we are not careful, and how to remedy that with fresh data and common sense.

Do not feel bad if you need to go back and review some of the content in this chapter, because there's a lot to digest. It is also important to get in the hypothesis-testing mindset if you want to be successful at a data science and machine learning career. Few practitioners take time to link statistics and hypothesis-testing concepts to machine learning, and that is unfortunate.

Understandability and explainability is the next frontier of machine learning, so continue to learn and integrate these ideas as you progress throughout the rest of this book and the rest of your career.

Exercises

1. You bought a spool of 1.75 mm filament for your 3D printer. You want to measure how close the filament diameter really is to 1.75 mm. You use a caliper tool and sample the diameter five times on the spool:

   1.78, 1.75, 1.72, 1.74, 1.77

   Calculate the mean and standard deviation for this set of values.

2. A manufacturer says the Z-Phone smart phone has a mean consumer life of 42 months with a standard deviation of 8 months. Assuming a normal distribution, what is the probability a given random Z-Phone will last between 20 and 30 months?

3. I am skeptical that my 3D printer filament is not 1.75 mm in average diameter as advertised. I sampled 34 measurements with my tool. The sample mean is 1.715588 and the sample standard deviation is 0.029252.

   What is the 99% confidence interval for the mean of my entire spool of filament?

4. Your marketing department has started a new advertising campaign and wants to know if it affected sales, which in the past averaged $10,345 a day with a standard

deviation of $552. The new advertising campaign ran for 45 days and averaged $11,641 in sales.

Did the campaign affect sales? Why or why not? (Use a two-tailed test for more reliable significance.)

Answers are in Appendix B.

## CHAPTER 4

Linear Algebra

Changing gears a little bit, let's venture away from probability and statistics and into linear algebra. Sometimes people confuse linear algebra with basic algebra, thinking maybe it has to do with plotting lines using the algebraic function $y = mx + b$. This is why linear algebra probably should have been called "vector algebra" or "matrix algebra" because it is much more abstract. Linear systems play a role but in a much more metaphysical way.

So, what exactly is linear algebra? Well, *linear algebra* concerns itself with linear systems but represents them through vector spaces and matrices. If you do not know what a vector or a matrix is, do not worry! We will define and explore them in depth. Linear algebra is hugely fundamental to many applied areas of math, statistics, operations research, data science, and machine learning. When you work with data in any of these areas, you are using linear algebra and perhaps you may not even know it.

You can get away with not learning linear algebra for a while, using machine learning and statistics libraries that do it all for you. But if you are going to get intuition behind these black boxes and be more effective at working with data, understanding the fundamentals of linear algebra is inevitable. Linear algebra is an enormous topic that can fill thick textbooks, so of course we cannot gain total mastery in just one chapter of this book. However, we can learn enough to be more comfortable with it and navigate the data science domain effectively. There will also be opportunities to apply it in the remaining chapters in this book, including Chapters 5 and 7.

What Is a Vector?

Simply put, a *vector* is an arrow in space with a specific direction and length, often representing a piece of data. It is the central building block of linear algebra, including matrices and linear transformations. In its fundamental form, it has no concept of location so always imagine its tail starts at the origin of a Cartesian plane (0,0).

Figure 4-1 shows a vector $\vec{v}$ that moves three steps in the horizontal direction and two steps in the vertical direction.

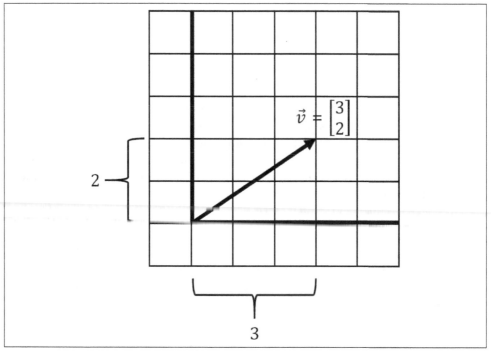

Figure 4-1. A simple vector

To emphasize again, the purpose of the vector is to visually represent a piece of data. If you have a data record for the square footage of a house 18,000 square feet and its valuation $260,000, we could express that as a vector [18000, 260000], stepping 18,000 steps in the horizontal direction and 260,000 steps in the vertical direction.

We declare a vector mathematically like this:

$$\vec{v} = \begin{bmatrix} x \\ y \end{bmatrix}$$

$$\vec{v} = \begin{bmatrix} 3 \\ 2 \end{bmatrix}$$

We can declare a vector using a simple Python collection, like a Python list as shown in Example 4-1.

Example 4-1. Declaring a vector in Python using a list

```
v = [3, 2]
print(v)
```

However, when we start doing mathematical computations with vectors, especially when doing tasks like machine learning, we should probably use the NumPy library as it is more efficient than plain Python. You can also use SymPy to perform linear algebra operations, and we will use it occasionally in this chapter when decimals become inconvenient. However, NumPy is what you will likely use in practice so that is what we will mainly stick to.

To declare a vector, you can use NumPy's `array()` function and then can pass a collection of numbers to it as shown in Example 4-2.

Example 4-2. Declaring a vector in Python using NumPy

```
import numpy as np
v = np.array([3, 2])
print(v)
```

Python Is Slow, Its Numerical Libraries Are Not

Python is a computationally slow language platform, as it does not compile to lower-level machine code and bytecode like Java, C#, C, etc. It is dynamically interpreted at runtime. However, Python's numeric and scientific libraries are not slow. Libraries like NumPy are typically written in low-level languages like C and C++, hence why they are computationally efficient. Python really acts as "glue code" integrating these libraries for your tasks.

A vector has countless practical applications. In physics, a vector is often thought of as a direction and magnitude. In math, it is a direction and scale on an XY plane, kind of like a movement. In computer science, it is an array of numbers storing data. The computer science context is the one we will become the most familiar with as data science professionals. However, it is important we never forget the visual aspect so we do not think of vectors as esoteric grids of numbers. Without a visual understanding, it is almost impossible to grasp many fundamental linear algebra concepts like linear dependence and determinants.

Here are some more examples of vectors. In Figure 4-2 note that some of these vectors have negative directions on the X and Y scales. Vectors with negative directions will have an impact when we combine them later, essentially subtracting rather than adding them together.

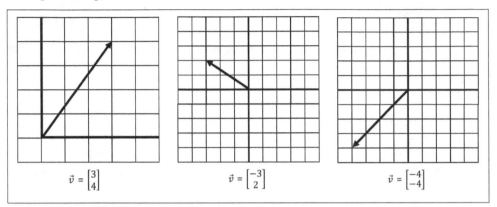

Figure 4-2. A sampling of different vectors

Why Are Vectors Useful?

A struggle many people have with vectors (and linear algebra in general) is understanding why they are useful. They are a highly abstract concept but they have many tangible applications. Computer graphics are easier to work with when you have a grasp of vectors and linear transformations (the fantastic Manim visualization library (*https://oreil.ly/Os5WK*) defines animations and transformations with vectors). When doing statistics and machine learning work, data is often imported and turned into numerical vectors so we can work with it. Solvers like the one in Excel or Python PuLP use linear programming, which uses vectors to maximize a solution while meeting those constraints. Even video games and flight simulators use vectors and linear algebra to model not just graphics but also physics. I think what makes vectors so hard to grasp is not that their application is unobvious, but rather these applications are so diverse it is hard to see the generalization.

Note also vectors can exist on more than two dimensions. Next we declare a three-dimensional vector along axes x, y, and z:

$$\vec{v} = \begin{bmatrix} x \\ y \\ z \end{bmatrix} = \begin{bmatrix} 4 \\ 1 \\ 2 \end{bmatrix}$$

To create this vector, we are stepping four steps in the x direction, one in the y direction, and two in the z direction. Here it is visualized in Figure 4-3. Note that we no longer are showing a vector on a two-dimensional grid but rather a three-dimensional space with three axes: x, y, and z.

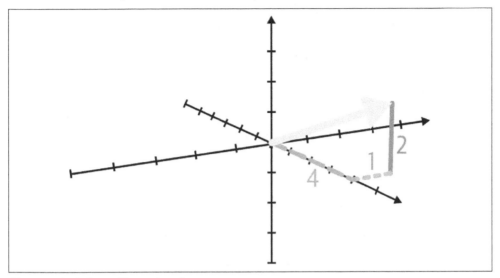

Figure 4-3. A three-dimensional vector

Naturally, we can express this three-dimensional vector in Python using three numeric values, as declared in Example 4-3.

Example 4-3. Declaring a three-dimensional vector in Python using NumPy

```
import numpy as np
v = np.array([4, 1, 2])
print(v)
```

Like many mathematical models, visualizing more than three dimensions is challenging and something we will not expend energy doing in this book. But numerically, it is still straightforward. Example 4-4 shows how we declare a five-dimensional vector mathematically in Python.

$$\vec{v} = \begin{bmatrix} 6 \\ 1 \\ 5 \\ 8 \\ 3 \end{bmatrix}$$

Example 4-4. Declaring a five-dimensional vector in Python using NumPy

```
import numpy as np
v = np.array([6, 1, 5, 8, 3])
print(v)
```

Adding and Combining Vectors

On their own, vectors are not terribly interesting. You express a direction and size, kind of like a movement in space. But when you start combining vectors, known as *vector addition*, it starts to get interesting. We effectively combine the movements of two vectors into a single vector.

Say we have two vectors $\vec{v}$ and $\vec{w}$ as shown in Figure 4-4. How do we add these two vectors together?

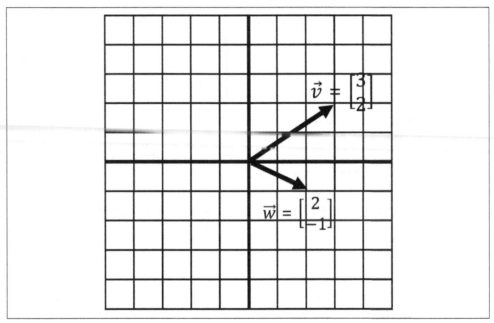

Figure 4-4. Adding two vectors together

We will get to why adding vectors is useful in a moment. But if we wanted to combine these two vectors, including their direction and scale, what would that look like? Numerically, this is straightforward. You simply add the respective x-values and then the y-values into a new vector as shown in Example 4-5.

$$\vec{v} = \begin{bmatrix} 3 \\ 2 \end{bmatrix}$$

$$\vec{w} = \begin{bmatrix} 2 \\ -1 \end{bmatrix}$$

$$\vec{v} + \vec{w} = \begin{bmatrix} 3 + 2 \\ 2 + -1 \end{bmatrix} = \begin{bmatrix} 5 \\ 1 \end{bmatrix}$$

Example 4-5. Adding two vectors in Python using NumPy

```
from numpy import array

v = array([3,2])
w = array([2,-1])

# sum the vectors
v_plus_w = v + w

# display summed vector
print(v_plus_w) # [5, 1]
```

But what does this mean visually? To visually add these two vectors together, connect one vector after the other and walk to the tip of the last vector (Figure 4-5). The point you end at is a new vector, the result of summing the two vectors.

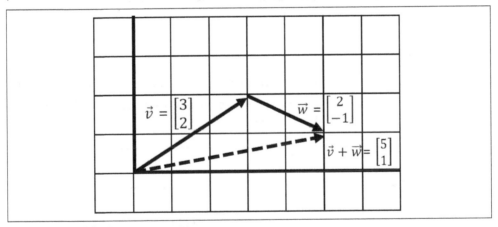

Figure 4-5. Adding two vectors into a new vector

As seen in Figure 4-5, when we walk to the end of the last vector $\vec{w}$ we end up with a new vector [5, 1]. This new vector is the result of summing $\vec{v}$ and $\vec{w}$. In practice, this can be simply adding data together. If we were totaling housing values and their square footage in an area, we would be adding several vectors into a single vector in this manner.

Note that it does not matter whether we add $\vec{v}$ before $\vec{w}$ or vice versa, which means it is *commutative* and order of operation does not matter. If we walk $\vec{w}$ before $\vec{v}$ we end up with the same resulting vector [5, 1] as visualized in Figure 4-6.

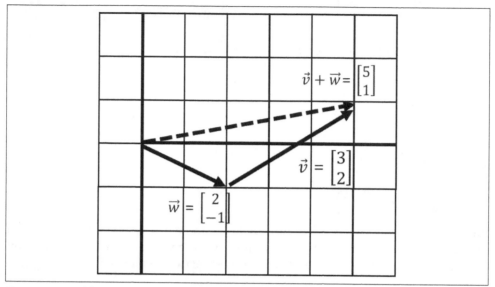

Figure 4-6. Adding vectors is commutative

Scaling Vectors

Scaling is growing or shrinking a vector's length. You can grow/shrink a vector by multiplying or scaling it with a single value, known as a *scalar*. Figure 4-7 is vector $\vec{v}$ being scaled by a factor of 2, which doubles it.

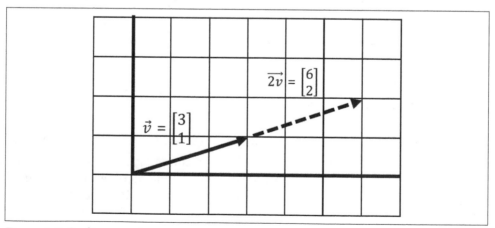

Figure 4-7. Scaling a vector

Mathematically, you multiply each element of the vector by the scalar value:

$$\vec{v} = \begin{bmatrix} 3 \\ 1 \end{bmatrix}$$

$$2\vec{v} = 2\begin{bmatrix} 3 \\ 1 \end{bmatrix} = \begin{bmatrix} 3 \times 2 \\ 1 \times 2 \end{bmatrix} = \begin{bmatrix} 6 \\ 2 \end{bmatrix}$$

Performing this scaling operation in Python is as easy as multiplying a vector by the scalar, as coded in Example 4-6.

Example 4-6. Scaling a number in Python using NumPy

```
from numpy import array

v = array([3,1])

# scale the vector
scaled_v = 2.0 * v

# display scaled vector
print(scaled_v) # [6 2]
```

Here in Figure 4-8 $\vec{v}$ is being scaled down by factor of .5, which halves it.

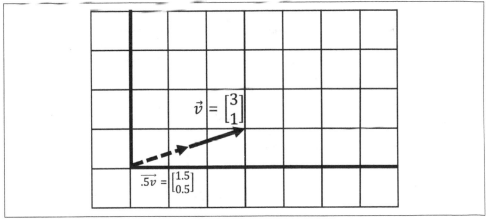

Figure 4-8. Scaling down a vector by half

Manipulating Data Is Manipulating Vectors

Every data operation can be thought of in terms of vectors, even simple averages.

Take scaling, for example. Let's say we were trying to get the average house value and average square footage for an entire neighborhood. We would add the vectors together to combine their value and square footage respectively, giving us one giant vector containing both total value and total square footage. We then scale down the vector by dividing by the number of houses N, which really is multiplying by $1/N$. We now have a vector containing the average house value and average square footage.

An important detail to note here is that scaling a vector does not change its direction, only its magnitude. But there is one slight exception to this rule as visualized in Figure 4-9. When you multiply a vector by a negative number, it flips the direction of the vector as shown in the image.

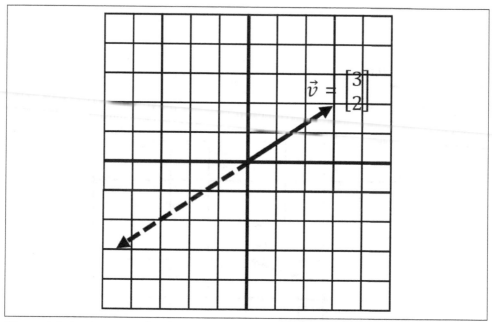

Figure 4-9. A negative scalar flips the vector direction

When you think about it, though, scaling by a negative number has not really changed direction in that it still exists on the same line. This segues to a key concept called linear dependence.

Span and Linear Dependence

These two operations, adding two vectors and scaling them, brings about a simple but powerful idea. With these two operations, we can combine two vectors and scale them to create any resulting vector we want. Figure 4-10 shows six examples of taking two vectors $\vec{v}$ and $\vec{w}$, and scaling and combining. These vectors $\vec{v}$ and $\vec{w}$, fixed in two different directions, can be scaled and added to create *any* new vector $\overrightarrow{v + w}$.

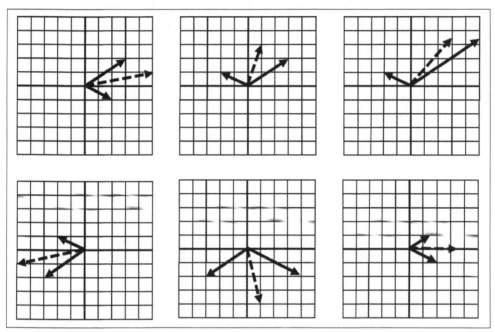

Figure 4-10. Scaling two added vectors allows us to create any new vector

Again, $\vec{v}$ and $\vec{w}$ are fixed in direction, except for flipping with negative scalars, but we can use scaling to freely create any vector composed of $\overrightarrow{v + w}$.

This whole space of possible vectors is called *span*, and in most cases our span can create unlimited vectors off those two vectors, simply by scaling and summing them. When we have two vectors in two different directions, they are *linearly independent* and have this unlimited span.

But in what case are we limited in the vectors we can create? Think about it and read on.

What happens when two vectors exist in the same direction, or exist on the same line? The combination of those vectors is also stuck on the same line, limiting our span to just that line. No matter how you scale it, the resulting sum vector is also stuck on that same line. This makes them *linearly dependent*, as shown in Figure 4-11.

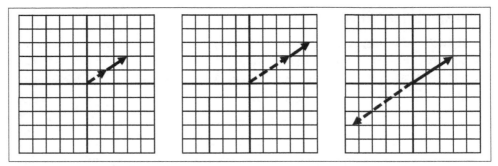

Figure 4-11. Linearly dependent vectors

The span here is stuck on the same line as the two vectors it is made out of. Because the two vectors exist on the same underlying line, we cannot flexibly create any new vector through scaling.

In three or more dimensions, when we have a linearly dependent set of vectors, we often get stuck on a plane in a smaller number of dimensions. Here is an example of being stuck on a two-dimensional plane even though we have three-dimensional vectors as declared in Figure 4-12.

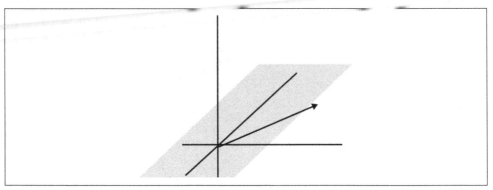

Figure 4-12. Linear dependence in three dimensions; note our span is limited to a flat plane

Later we will learn a simple tool called the determinant to check for linear dependence, but why do we care whether two vectors are linearly dependent or independent? A lot of problems become difficult or unsolvable when they are linearly dependent. For example, when we learn about systems of equations later in this chapter, a linearly dependent set of equations can cause variables to disappear and make the problem unsolvable. But if you have linear independence, that flexibility to create any vector you need from two or more vectors becomes invaluable to solve for a solution!

Linear Transformations

This concept of adding two vectors with fixed direction, but scaling them to get different combined vectors, is hugely important. This combined vector, except in cases of linear dependence, can point in any direction and have any length we choose. This sets up an intuition for linear transformations where we use a vector to transform another vector in a function-like manner.

Basis Vectors

Imagine we have two simple vectors $\hat{i}$ and $\hat{j}$ ("i-hat" and "j-hat"). These are known as *basis vectors*, which are used to describe transformations on other vectors. They typically have a length of 1 and point in perpendicular positive directions as visualized in Figure 4-13.

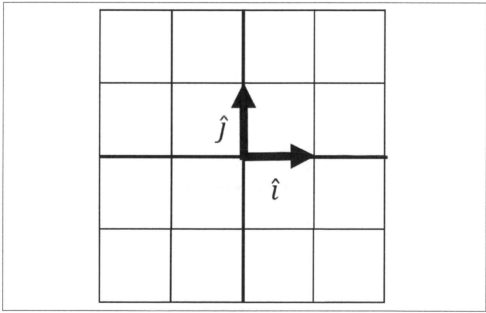

Figure 4-13. Basis vectors $\hat{i}$ and $\hat{j}$

Think of the basis vectors as building blocks to build or transform any vector. Our basis vector is expressed in a 2 × 2 matrix, where the first column is $\hat{i}$ and the second column is $\hat{j}$:

$$\hat{i} = \begin{bmatrix} 1 \\ 0 \end{bmatrix}$$

$$\hat{j} = \begin{bmatrix} 0 \\ 1 \end{bmatrix}$$

$$\text{basis} = \begin{bmatrix} 1 & 0 \\ 0 & 1 \end{bmatrix}$$

A *matrix* is a collection of vectors (such as $\hat{i}$, $\hat{j}$) that can have multiple rows and columns and is a convenient way to package data. We can use $\hat{i}$ and $\hat{j}$ to create any vector we want by scaling and adding them. Let's start with each having a length of 1 and showing the resulting vector $\vec{v}$ in Figure 4-14.

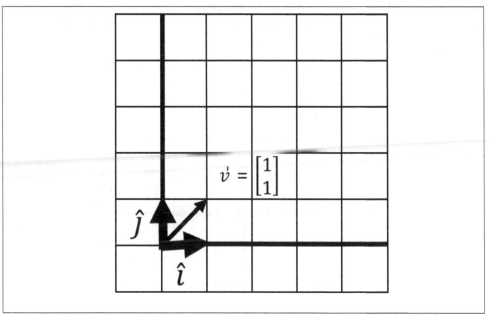

Figure 4-14. Creating a vector from basis vectors

I want vector $\vec{v}$ to land at [3, 2]. What happens to $\vec{v}$ if we stretch $\hat{i}$ by a factor of 3 and $\hat{j}$ by a factor of 2? First we scale them individually as shown here:

$$3\hat{i} = 3\begin{bmatrix} 1 \\ 0 \end{bmatrix} = \begin{bmatrix} 3 \\ 0 \end{bmatrix}$$

$$2\hat{j} = 2\begin{bmatrix} 0 \\ 1 \end{bmatrix} = \begin{bmatrix} 0 \\ 2 \end{bmatrix}$$

If we stretched space in these two directions, what does this do to $\vec{v}$? Well, it is going to stretch with $\hat{i}$ and $\hat{j}$. This is known as a *linear transformation*, where we transform a vector with stretching, squishing, sheering, or rotating by tracking basis vector movements. In this case (Figure 4-15), scaling $\hat{i}$ and $\hat{j}$ has stretched space along with our vector $\vec{v}$.

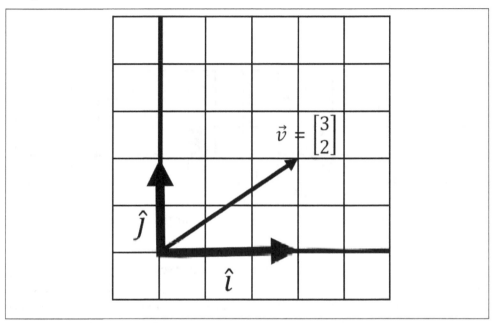

Figure 4-15. A linear transformation

But where does $\vec{v}$ land? It is easy to see where it lands here, which is [3, 2]. Recall that vector $\vec{v}$ is composed of adding $\hat{i}$ and $\hat{j}$. So we simply take the stretched $\hat{i}$ and $\hat{j}$ and add them together to see where vector $\vec{v}$ has landed:

$$\vec{v}_{new} = \begin{bmatrix} 3 \\ 0 \end{bmatrix} + \begin{bmatrix} 0 \\ 2 \end{bmatrix} = \begin{bmatrix} 3 \\ 2 \end{bmatrix}$$

Generally, with linear transformations, there are four movements you can achieve, as shown in Figure 4-16.

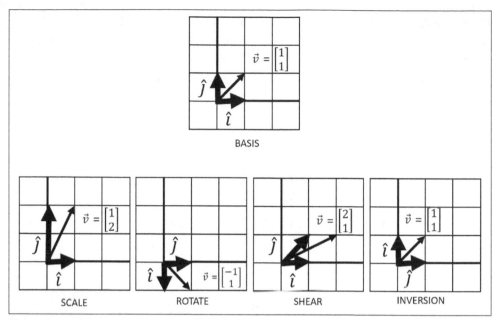

Figure 4-16. Four movements can be achieved with linear transformations

These four linear transformations are a central part of linear algebra. Scaling a vector will stretch or squeeze it. Rotations will turn the vector space, and inversions will flip the vector space so that $\hat{i}$ and $\hat{j}$ swap respective places. A shear is easier to describe visually, but it displaces each point in a fixed direction proportionally to its distance from a given line parallel to that direction.

It is important to note that you cannot have transformations that are nonlinear, resulting in curvy or squiggly transformations that no longer respect a straight line. This is why we call it linear algebra, not nonlinear algebra!

Matrix Vector Multiplication

This brings us to our next big idea in linear algebra. This concept of tracking where $\hat{i}$ and $\hat{j}$ land after a transformation is important because it allows us not just to create vectors but also to transform existing vectors. If you want true linear algebra enlightenment, think why creating vectors and transforming vectors are actually the same thing. It's all a matter of relativity given your basis vectors being a starting point before and after a transformation.

The formula to transform a vector $\vec{v}$ given basis vectors $\hat{i}$ and $\hat{j}$ packaged as a matrix is:

$$\begin{bmatrix} x_{new} \\ y_{new} \end{bmatrix} = \begin{bmatrix} a & b \\ c & d \end{bmatrix} \begin{bmatrix} x \\ y \end{bmatrix}$$

$$\begin{bmatrix} x_{new} \\ y_{new} \end{bmatrix} = \begin{bmatrix} ax + by \\ cx + dy \end{bmatrix}$$

$\widehat{i}$ is the first column $[a, c]$ and $\widehat{j}$ is the column $[b, d]$. We package both of these basis vectors as a matrix, which again is a collection of vectors expressed as a grid of numbers in two or more dimensions. This transformation of a vector by applying basis vectors is known as *matrix vector multiplication*. This may seem contrived at first, but this formula is a shortcut for scaling and adding $\widehat{i}$ and $\widehat{j}$ just like we did earlier adding two vectors, and applying the transformation to any vector $\overrightarrow{v}$.

So in effect, a matrix really is a transformation expressed as basis vectors.

To execute this transformation in Python using NumPy, we will need to declare our basis vectors as a matrix and then apply it to vector $\overrightarrow{v}$ using the dot() operator (Example 4-7). The dot() operator will perform this scaling and addition between our matrix and vector as we just described. This is known as the *dot product*, and we will explore it throughout this chapter.

Example 4-7. Matrix vector multiplication in NumPy

```
from numpy import array

# compose basis matrix with i-hat and j-hat
basis = array(
    [[3, 0],
     [0, 2]]
 )

# declare vector v
v = array([1,1])

# create new vector
# by transforming v with dot product
new_v = basis.dot(v)

print(new_v) # [3, 2]
```

When thinking in terms of basis vectors, I prefer to break out the basis vectors and then compose them together into a matrix. Just note you will need to *transpose*, or swap the columns and rows. This is because NumPy's array() function will do the opposite orientation we want, populating each vector as a row rather than a column. Transposition in NumPy is demonstrated in Example 4-8.

Example 4-8. Separating the basis vectors and applying them as a transformation

```
from numpy import array

# Declare i-hat and j-hat
i_hat = array([2, 0])
j_hat = array([0, 3])

# compose basis matrix using i-hat and j-hat
# also need to transpose rows into columns
basis = array([i_hat, j_hat]).transpose()

# declare vector v
v = array([1,1])

# create new vector
# by transforming v with dot product
new_v = basis.dot(v)

print(new_v) # [2, 3]
```

Here's another example. Let's start with vector $\vec{v}$ being [2, 1] and $\hat{i}$ and $\hat{j}$ start at [1, 0] and [0, 1], respectively. We then transform $\hat{i}$ and $\hat{j}$ to [2, 0] and [0, 3]. What happens to vector $\vec{v}$? Working this out mathematically by hand using our formula, we get this:

$$\begin{bmatrix} x_{new} \\ y_{new} \end{bmatrix} = \begin{bmatrix} a & b \\ c & d \end{bmatrix}\begin{bmatrix} x \\ y \end{bmatrix} = \begin{bmatrix} ax + by \\ cx + dy \end{bmatrix}$$

$$\begin{bmatrix} x_{new} \\ y_{new} \end{bmatrix} = \begin{bmatrix} 2 & 0 \\ 0 & 3 \end{bmatrix}\begin{bmatrix} 2 \\ 1 \end{bmatrix} = \begin{bmatrix} (2)(2) + (0)(1) \\ (2)(0) + (3)(1) \end{bmatrix} = \begin{bmatrix} 4 \\ 3 \end{bmatrix}$$

Example 4-9 shows this solution in Python.

Example 4-9. Transforming a vector using NumPy

```
from numpy import array

# Declare i-hat and j-hat
i_hat = array([2, 0])
j_hat = array([0, 3])

# compose basis matrix using i-hat and j-hat
# also need to transpose rows into columns
basis = array([i_hat, j_hat]).transpose()
```

```
# declare vector v 0
v = array([2,1])

# create new vector
# by transforming v with dot product
new_v = basis.dot(v)

print(new_v) # [4, 3]
```

The vector $\vec{v}$ now lands at [4, 3]. Figure 4-17 shows what this transformation looks like.

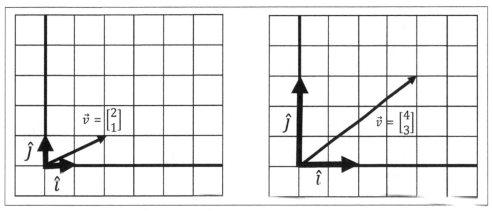

Figure 4-17. A stretching linear transformation

Here is an example that jumps things up a notch. Let's take vector $\vec{v}$ of value [2, 1]. $\hat{i}$ and $\hat{j}$ start at [1, 0] and [0, 1], but then are transformed and land at [2, 3] and [2, -1]. What happens to $\vec{v}$? Let's look in Figure 4-18 and Example 4-10.

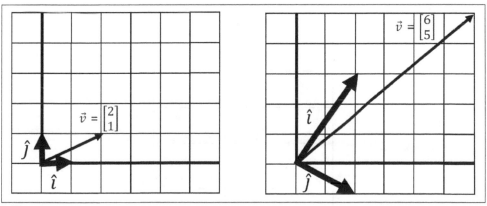

Figure 4-18. A linear transformation that does a rotation, shear, and flipping of space

Example 4-10. A more complicated transformation

```
from numpy import array

# Declare i-hat and j-hat
i_hat = array([2, 3])
j_hat = array([2, -1])

# compose basis matrix using i-hat and j-hat
# also need to transpose rows into columns
basis = array([i_hat, j_hat]).transpose()

# declare vector v 0
v = array([2,1])

# create new vector
# by transforming v with dot product
new_v = basis.dot(v)

print(new_v) # [6, 5]
```

A lot has happened here. Not only did we scale $\hat{i}$ and $\hat{j}$ and elongate vector $\vec{v}$. We actually sheared, rotated, and flipped space, too. You know space was flipped when $\hat{i}$ and $\hat{j}$ change places in their clockwise orientation, and we will learn how to detect this with determinants later in this chapter.

Basis Vectors in 3D and Beyond

You might be wondering how we think of vector transformations in three dimensions or more. The concept of basis vectors extends quite nicely. If I have a three-dimensional vector space, then I have basis vectors $\hat{i}$, $\hat{j}$, and $\hat{k}$. I just keep adding more letters from the alphabet for each new dimension (Figure 4-19).

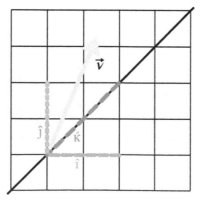

Figure 4-19. 3D basis vector

Something else that is worth pointing out is some linear transformations can shift a vector space into fewer or more dimensions, and that is exactly what nonsquare matrices will do (where number of rows and columns are not equal). In the interest of time we cannot explore this. But 3Blue1Brown explains and animates this concept beautifully (*https://oreil.ly/TsoSJ*).

Matrix Multiplication

We learned how to multiply a vector and a matrix, but what exactly does multiplying two matrices accomplish? Think of *matrix multiplication* as applying multiple transformations to a vector space. Each transformation is like a function, where we apply the innermost first and then apply each subsequent transformation outward.

Here is how we apply a rotation and then a shear to any vector $\vec{v}$ with value $[x, y]$:

$$\begin{bmatrix} 1 & 1 \\ 0 & 1 \end{bmatrix} \begin{bmatrix} 0 & -1 \\ 1 & 0 \end{bmatrix} \begin{bmatrix} x \\ y \end{bmatrix}$$

We can actually consolidate these two transformations by using this formula, applying one transformation onto the last. You multiply and add each row from the first matrix to each respective column of the second matrix, in an "over and-down! over-and-down!" pattern:

$$\begin{bmatrix} a & b \\ c & d \end{bmatrix} \begin{bmatrix} e & f \\ g & h \end{bmatrix} = \begin{bmatrix} ae + bg & af + bh \\ ce + dg & cf + dh \end{bmatrix}$$

So we can actually consolidate these two separate transformations (rotation and shear) into a single transformation:

$$\begin{bmatrix} 1 & 1 \\ 0 & 1 \end{bmatrix} \begin{bmatrix} 0 & -1 \\ 1 & 0 \end{bmatrix} \begin{bmatrix} x \\ y \end{bmatrix}$$

$$= \begin{bmatrix} (1)(0) + (1)(1) & (-1)(1) + (1)(0) \\ (0)(0) + (1)(1) & (0)(-1) + (1)(0) \end{bmatrix} \begin{bmatrix} x \\ y \end{bmatrix}$$

$$= \begin{bmatrix} 1 & -1 \\ 1 & 0 \end{bmatrix} \begin{bmatrix} x \\ y \end{bmatrix}$$

To execute this in Python using NumPy, you can combine the two matrices simply using the `matmul()` or @ operator (Example 4-11). We will then turn around and use this consolidated tranformation and apply it to a vector $[1, 2]$.

Example 4-11. Combining two transformations

```
from numpy import array

# Transformation 1
i_hat1 = array([0, 1])
j_hat1 = array([-1, 0])
transform1 = array([i_hat1, j_hat1]).transpose()

# Transformation 2
i_hat2 = array([1, 0])
j_hat2 = array([1, 1])
transform2 = array([i_hat2, j_hat2]).transpose()

# Combine Transformations
combined = transform2 @ transform1

# Test
print("COMBINED MATRIX:\n {}".format(combined))

v = array([1, 2])
print(combined.dot(v))  # [-1, 1]
```

Using dot() Versus matmul() and @

In general, you want to prefer matmul() and its shorthand @ to combine matrices rather than the dot() operator in NumPy. The former generally has a preferable policy for higher-dimensional matrices and how the elements are broadcasted.

If you like diving into these kinds of implementation details, this StackOverflow question is a good place to start (*https://oreil.ly/ YX83Q*).

Note that we also could have applied each transformation individually to vector $\vec{v}$ and still have gotten the same result. If you replace the last line with these three lines applying each transformation, you will still get [-1, 1] on that new vector:

```
rotated = transform1.dot(v)
sheared = transform2.dot(rotated)
print(sheared) # [-1, 1]
```

Note that the order you apply each transformation matters! If we apply transformation1 on transformation2, we get a different result of [-2, 3] as calculated in Example 4-12. So matrix dot products are not commutative, meaning you cannot flip the order and expect the same result!

Example 4-12. Applying the transformations in reverse

```
from numpy import array

# Transformation 1
i_hat1 = array([0, 1])
j_hat1 = array([-1, 0])
transform1 = array([i_hat1, j_hat1]).transpose()

# Transformation 2
i_hat2 = array([1, 0])
j_hat2 = array([1, 1])
transform2 = array([i_hat2, j_hat2]).transpose()

# Combine Transformations, apply sheer first and then rotation
combined = transform1 @ transform2

# Test
print("COMBINED MATRIX:\n {}".format(combined))

v = array([1, 2])
print(combined.dot(v)) # [-2, 3]
```

Think of each transformation as a function, and we apply them from the innermost to outermost just like nested function calls.

Linear Transformations in Practice

You might be wondering what all these linear transformations and matrices have to do with data science and machine learning. The answer is everything! From importing data to numerical operations with linear regression, logistic regression, and neural networks, linear transformations are the heart of mathematically manipulated data.

In practice, however, you will rarely take the time to geometrically visualize your data as vector spaces and linear transformations. You will be dealing with too many dimensions to do this productively. But it is good to be mindful of of the geometric interpretation just to understand what these contrived-looking numerical operations do! Otherwise, you are just memorizing numerical operation patterns without any context. It also makes new linear algebra concepts like determinants more obvious.

Determinants

When we perform linear transformations, we sometimes "expand" or "squish" space and the degree this happens can be helpful. Take a sampled area from the vector space in Figure 4-20: what happens to it after we scale $\hat{i}$ and $\hat{j}$?

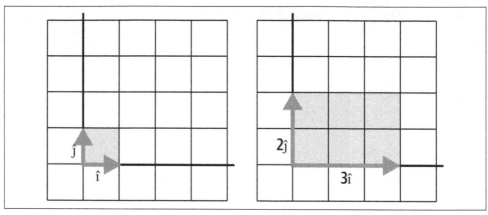

Figure 4-20. A determinant measures how a linear transformation scales an area

Note it increases in area by a factor of 6.0, and this factor is known as a *determinant*. Determinants describe how much a sampled area in a vector space changes in scale with linear transformations, and this can provide helpful information about the transformation.

Example 4-13 shows how to calculate this determinant in Python.

Example 4-13. Calculating a determinant

```
from numpy.linalg import det
from numpy import array

i_hat = array([3, 0])
j_hat = array([0, 2])

basis = array([i_hat, j_hat]).transpose()

determinant = det(basis)

print(determinant) # prints 6.0
```

Simple shears and rotations should not affect the determinant, as the area will not change. Figure 4-21 and Example 4-14 shows a simple shear and the determinant remains a factor 1.0, showing it is unchanged.

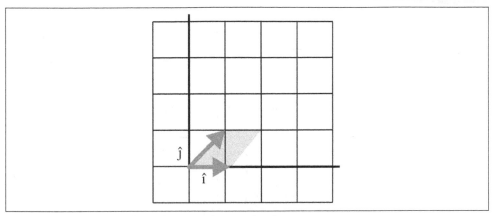

Figure 4-21. A simple shear does not change the determinant

Example 4-14. A determinant for a shear

```
from numpy.linalg import det
from numpy import array

i_hat = array([1, 0])
j_hat = array([1, 1])

basis = array([i_hat, j_hat]).transpose()

determinant = det(basis)

print(determinant) # prints 1.0
```

But scaling will increase or decrease the determinant, as that will increase/decrease the sampled area. When the orientation flips $(\hat{i}, \hat{j}$ swap clockwise positions), then the determinant will be negative. Figure 4-22 and Example 4-15 illustrate a determinant showing a transformation that not only scaled but also flipped the orientation of the vector space.

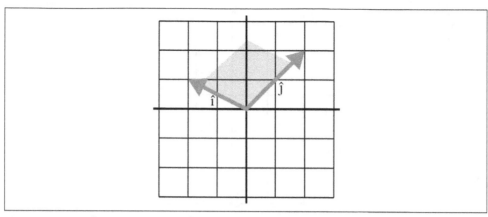

Figure 4-22. A determinant on a flipped space is negative

Example 4-15. A negative determinant

```
from numpy.linalg import det
from numpy import array

i_hat = array([-2, 1])
j_hat = array([1, 2])

basis = array([i_hat, j_hat]).transpose()

determinant = det(basis)

print(determinant) # prints -5.0
```

Because this determinant is negative, we quickly see that the orientation has flipped. But by far the most critical piece of information the determinant tells you is whether the transformation is linearly dependent. If you have a determinant of 0, that means all of the space has been squished into a lesser dimension.

In Figure 4-23 we see two linearly dependent transformations, where a 2D space is compressed into one dimension and a 3D space is compressed into two dimensions. The area and volume respectively in both cases are 0!

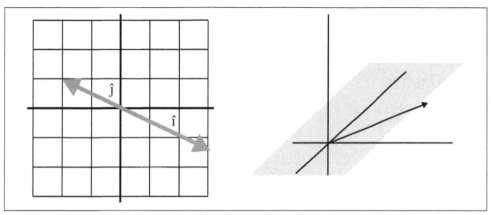

Figure 4-23. Linear dependence in 2D and 3D

Example 4-16 shows the code for the preceding 2D example squishing an entire 2D space into a single one-dimensional number line.

Example 4-16. A determinant of zero

```
from numpy.linalg import det
from numpy import array

i_hat = array([3, -1.5])
j_hat = array([-2, 1])

basis = array([i_hat, j_hat]).transpose()

determinant = det(basis)

print(determinant) # prints 0.0
```

So testing for a 0 determinant is highly helpful to determine if a transformation has linear dependence. When you encounter this you will likely find a difficult or unsolvable problem on your hands.

Special Types of Matrices

There are a few notable cases of matrices that we should cover.

Square Matrix

The *square matrix* is a matrix that has an equal number of rows and columns:

$$\begin{bmatrix} 4 & 2 & 7 \\ 5 & 1 & 9 \\ 4 & 0 & 1 \end{bmatrix}$$

They are primarily used to represent linear transformations and are a requirement for many operations like eigendecomposition.

Identity Matrix

The *identity matrix* is a square matrix that has a diagonal of 1s while the other values are 0:

$$\begin{bmatrix} 1 & 0 & 0 \\ 0 & 1 & 0 \\ 0 & 0 & 1 \end{bmatrix}$$

What's the big deal with identity matrices? Well, when you have an identity matrix, you essentially have undone a transformation and found your starting basis vectors. This will play a big role in solving systems of equations in the next section.

Inverse Matrix

An *inverse matrix* is a matrix that undoes the transformation of another matrix. Let's say I have matrix A:

$$A = \begin{bmatrix} 4 & 2 & 4 \\ 5 & 3 & 7 \\ 9 & 3 & 6 \end{bmatrix}$$

The inverse of matrix A is called A^{-1}. We will learn how to calculate the inverse using Sympy or NumPy in the next section, but this is what the inverse of matrix A is:

$$A^{-1} = \begin{bmatrix} -\dfrac{1}{2} & 0 & \dfrac{1}{3} \\ 5.5 & -2 & -\dfrac{4}{3} \\ -2 & 1 & \dfrac{1}{3} \end{bmatrix}$$

When we perform matrix multiplication between A^{-1} and A, we end up with an identity matrix. We will see this in action with NumPy and Sympy in the next section on systems of equations.

$$\begin{bmatrix} -\dfrac{1}{2} & 0 & \dfrac{1}{3} \\ 5.5 & -2 & -\dfrac{4}{3} \\ -2 & 1 & \dfrac{1}{3} \end{bmatrix} \begin{bmatrix} 4 & 2 & 4 \\ 5 & 3 & 7 \\ 9 & 3 & 6 \end{bmatrix} = \begin{bmatrix} 1 & 0 & 0 \\ 0 & 1 & 0 \\ 0 & 0 & 1 \end{bmatrix}$$

Diagonal Matrix

Similar to the identity matrix is the *diagonal matrix*, which has a diagonal of nonzero values while the rest of the values are 0. Diagonal matrices are desirable in certain computations because they represent simple scalars being applied to a vector space. It shows up in some linear algebra operations.

$$\begin{bmatrix} 4 & 0 & 0 \\ 0 & 2 & 0 \\ 0 & 0 & 5 \end{bmatrix}$$

Triangular Matrix

Similar to the diagonal matrix is the *triangular matrix*, which has a diagonal of nonzero values in front of a triangle of values, while the rest of the values are 0.

$$\begin{bmatrix} 4 & 2 & 9 \\ 0 & 1 & 6 \\ 0 & 0 & 5 \end{bmatrix}$$

Triangular matrices are desirable in many numerical analysis tasks, because they typically are easier to solve in systems of equations. They also show up in certain decomposition tasks like LU Decomposition (*https://oreil.ly/vYK8t*).

Sparse Matrix

Occasionally, you will run into matrices that are mostly zeroes and have very few nonzero elements. These are called *sparse matrices*. From a pure mathematical standpoint, they are not terribly interesting. But from a computing standpoint, they provide opportunities to create efficiency. If a matrix has mostly 0s, a sparse matrix implementation will not waste space storing a bunch of 0s, and instead only keep track of the cells that are nonzero.

$$\text{sparse:} \begin{bmatrix} 0 & 0 & 0 \\ 0 & 0 & 2 \\ 0 & 0 & 0 \\ 0 & 0 & 0 \end{bmatrix}$$

When you have large matrices that are sparse, you might explicitly use a sparse function to create your matrix.

Systems of Equations and Inverse Matrices

One of the basic use cases for linear algebra is solving systems of equations. It is also a good application to learn about inverse matrices. Let's say you are provided with the following equations and you need to solve for x, y, and z:

$$4x + 2y + 4z = 44$$
$$5x + 3y + 7z = 56$$
$$9x + 3y + 6z = 72$$

You can try manually experimenting with different algebraic operations to isolate the three variables, but if you want a computer to solve it you will need to express this problem in terms of matrices as shown next. Extract the coefficients into matrix A, the values on the right side of the equation into matrix B, and the unknown variables into matrix X:

$$A = \begin{bmatrix} 4 & 2 & 4 \\ 5 & 3 & 7 \\ 9 & 3 & 6 \end{bmatrix}$$

$$B = \begin{bmatrix} 44 \\ 56 \\ 72 \end{bmatrix}$$

$$X = \begin{bmatrix} x \\ y \\ z \end{bmatrix}$$

The function for a linear system of equations is $AX = B$. We need to transform matrix A with some other matrix X that will result in matrix B:

$AX = B$

$$\begin{bmatrix} 4 & 2 & 4 \\ 5 & 3 & 7 \\ 9 & 3 & 6 \end{bmatrix} \cdot \begin{bmatrix} x \\ y \\ z \end{bmatrix} = \begin{bmatrix} 44 \\ 56 \\ 72 \end{bmatrix}$$

We need to "undo" A so we can isolate X and get the values for x, y, and z. The way you undo A is to take the inverse of A denoted by A^{-1} and apply it to A via matrix multiplication. We can express this algebraically:

$AX = B$
$A^{-1}AX = A^{-1}B$
$X - 4^{-1}B$

To calculate the inverse of matrix A, we probably use a computer rather than searching for solutions by hand using Gaussian elimination, which we will not venture into in this book. Here is the inverse of matrix A:

$$A^{-1} = \begin{bmatrix} -\dfrac{1}{2} & 0 & \dfrac{1}{3} \\ 5.5 & -2 & \dfrac{4}{3} \\ -2 & 1 & \dfrac{1}{3} \end{bmatrix}$$

Note when we matrix multiply A^{-1} against A it will create an identity matrix, a matrix of all zeroes except for 1s in the diagonal. The identity matrix is the linear algebra equivalent of multiplying by 1, meaning it essentially has no effect and will effectively isolate values for x, y, and z:

$$A^{-1} = \begin{bmatrix} -\dfrac{1}{2} & 0 & \dfrac{1}{3} \\[2ex] 5.5 & -2 & \dfrac{4}{3} \\[2ex] -2 & 1 & \dfrac{1}{3} \end{bmatrix}$$

$$A = \begin{bmatrix} 4 & 2 & 4 \\ 5 & 3 & 7 \\ 9 & 3 & 6 \end{bmatrix}$$

$$A^{-1}A = \begin{bmatrix} 1 & 0 & 0 \\ 0 & 1 & 0 \\ 0 & 0 & 1 \end{bmatrix}$$

To see this identity matrix in action in Python, you will want to use SymPy instead of NumPy. The floating point decimals in NumPy will not make the identity matrix as obvious, but doing it symbolically in Example 4-17 we will see a clean, symbolic output. Note that to do matrix multiplication in SymPy we use the asterisk * rather than @.

Example 4-17. Using SymPy to study the inverse and identity matrix

```python
from sympy import *

# 4x + 2y + 4z = 44
# 5x + 3y + 7z = 56
# 9x + 3y + 6z = 72

A = Matrix([
    [4, 2, 4],
    [5, 3, 7],
    [9, 3, 6]
])

# dot product between A and its inverse
# will produce identity function
inverse = A.inv()
identity = inverse * A

# prints Matrix([[-1/2, 0, 1/3], [11/2, -2, -4/3], [-2, 1, 1/3]])
print("INVERSE: {}".format(inverse))

# prints Matrix([[1, 0, 0], [0, 1, 0], [0, 0, 1]])
print("IDENTITY: {}".format(identity))
```

In practice, the lack of floating point precision will not affect our answers too badly, so using NumPy should be fine to solve for *x*. Example 4-18 shows a solution with NumPy.

Example 4-18. Using NumPy to solve a system of equations

```
from numpy import array
from numpy.linalg import inv

# 4x + 2y + 4z = 44
# 5x + 3y + 7z = 56
# 9x + 3y + 6z = 72

A = array([
    [4, 2, 4],
    [5, 3, 7],
    [9, 3, 6]
])

B = array([
    44,
    56,
    72
])

X = inv(A).dot(B)

print(X) # [ 2. 34. -8.]
```

So $x = 2$, $y = 34$, and $z = -8$. Example 4-19 shows the full solution in SymPy as an alternative to NumPy.

Example 4-19. Using SymPy to solve a system of equations

```
from sympy import *

# 4x + 2y + 4z = 44
# 5x + 3y + 7z = 56
# 9x + 3y + 6z = 72

A = Matrix([
    [4, 2, 4],
    [5, 3, 7],
    [9, 3, 6]
])

B = Matrix([
    44,
    56,
    72
```

```
])

X = A.inv() * B

print(X) # Matrix([[2], [34], [-8]])
```

Here is the solution in mathematical notation:

$$A^{-1}B = X$$

$$\begin{bmatrix} -\dfrac{1}{2} & 0 & \dfrac{1}{3} \\ 5.5 & -2 & \dfrac{4}{3} \\ -2 & 1 & \dfrac{1}{3} \end{bmatrix} \begin{bmatrix} 44 \\ 56 \\ 72 \end{bmatrix} = \begin{bmatrix} x \\ y \\ z \end{bmatrix}$$

$$\begin{bmatrix} 2 \\ 34 \\ -8 \end{bmatrix} = \begin{bmatrix} x \\ y \\ z \end{bmatrix}$$

Hopefully, this gave you an intuition for inverse matrices and how they can be used to solve a system of equations.

Systems of Equations in Linear Programming

This method of solving systems of equations is used for linear programming as well, where inequalities define constraints and an objective is minimized/maximized.

PatrickJMT has a lot of good videos on Linear Programming (*https://bit.ly/3aVyrD6*). We also cover it briefly in Appendix A.

In practicality, you should rarely find it necessary to calculate inverse matrices by hand and can have a computer do it for you. But if you have a need or are curious, you will want to learn about Gaussian elimination. PatrickJMT on YouTube (*https://oreil.ly/RfXAv*) has a number of videos demonstrating Gaussian elimination.

Eigenvectors and Eigenvalues

Matrix decomposition is breaking up a matrix into its basic components, much like factoring numbers (e.g., 10 can be factored to 2×5).

Matrix decomposition is helpful for tasks like finding inverse matrices and calculating determinants, as well as linear regression. There are many ways to decompose a matrix depending on your task. In Chapter 5 we will use a matrix decomposition technique, QR decomposition, to perform a linear regression.

But in this chapter let's focus on a common method called eigendecomposition, which is often used for machine learning and principal component analysis. At this level we do not have the bandwidth to dive into each of these applications. For now, just know eigendecomposition is helpful for breaking up a matrix into components that are easier to work with in different machine learning tasks. Note also it only works on square matrices.

In eigendecomposition, there are two components: the eigenvalues denoted by lambda λ and eigenvector by v shown in Figure 4-24.

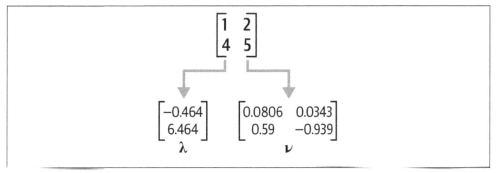

Figure 4-24. The eigenvector and eigenvalues

If we have a square matrix A, it has the following eigenvalue equation:

$$Av = \lambda v$$

If A is the original matrix, it is composed of eigenvector v and eigenvalue λ. There is one eigenvector and eigenvalue for each dimension of the parent matrix, and not all matrices can be decomposed into an eigenvector and eigenvalue. Sometimes complex (imaginary) numbers will even result.

Example 4-20 is how we calculate eigenvectors and eigenvalues in NumPy for a given matrix A.

Example 4-20. Performing eigendecomposition in NumPy

```
from numpy import array, diag
from numpy.linalg import eig, inv

A = array([
    [1, 2],
    [4, 5]
])
```

```
eigenvals, eigenvecs = eig(A)

print("EIGENVALUES")
print(eigenvals)
print("\nEIGENVECTORS")
print(eigenvecs)

"""
EIGENVALUES
[-0.46410162  6.46410162]

EIGENVECTORS
[[-0.80689822 -0.34372377]
 [ 0.59069049 -0.9390708 ]]
"""
```

So how do we rebuild matrix A from the eigenvectors and eigenvalues? Recall this formula:

$$Av = \lambda v$$

We need to make a few tweaks to the formula to reconstruct A:

$$A = Q \Lambda Q^{-1}$$

In this new formula, Q is the eigenvectors, Λ is the eigenvalues in diagonal form, and Q^{-1} is the inverse matrix of Q. Diagonal form means the vector is padded into a matrix of zeroes and occupies the diagonal line in a similar pattern to an identity matrix.

Example 4-21 brings the example full circle in Python, starting with decomposing the matrix and then recomposing it.

Example 4-21. Decomposing and recomposing a matrix in NumPy

```
from numpy import array, diag
from numpy.linalg import eig, inv

A = array([
    [1, 2],
    [4, 5]
])

eigenvals, eigenvecs = eig(A)

print("EIGENVALUES")
```

```
print(eigenvals)
print("\nEIGENVECTORS")
print(eigenvecs)

print("\nREBUILD MATRIX")
Q = eigenvecs
R = inv(Q)

L = diag(eigenvals)
B = Q @ L @ R

print(B)

"""
EIGENVALUES
[-0.46410162   6.46410162]

EIGENVECTORS
[[-0.80689822  -0.34372377]
 [ 0.59069049  -0.9390708 ]]

REBUILD MATRIX
[[1. 2.]
 [4. 5.]]
"""
```

As you can see, the matrix we rebuilt is the one we started with.

Conclusion

Linear algebra can be maddeningly abstract and it is full of mysteries and ideas to ponder. You may find the whole topic is one big rabbit hole, and you would be right! However, it is a good idea to continue being curious about it if you want to have a long, successful data science career. It is the foundation for statistical computing, machine learning, and other applied data science areas. Ultimately, it is the foundation for computer science in general. You can certainly get away with not knowing it for a while but you will encounter limitations in your understanding at some point.

You may wonder how these ideas are practical as they may feel theoretical. Do not worry; we will see some practical applications throughout this book. But the theory and geometric interpretations are important to have intuition when you work with data, and by understanding linear transformations visually you are prepared to take on more advanced concepts that may be thrown at you later in your pursuits.

If you want to learn more about linear programming, there is no better place than 3Blue1Brown's YouTube playlist "Essence of Linear Algebra" (*https://oreil.ly/FSCNz*). The linear algebra videos from PatrickJMT (*https://oreil.ly/Hx9GP*) are helpful as well.

If you want to get more comfortable with NumPy, the O'Reilly book *Python for Data Analysis* (2nd edition) by Wes McKinney is a recommended read. It does not focus much on linear algebra, but it does provide practical instruction on using NumPy, Pandas, and Python on datasets.

Exercises

1. Vector $\vec{v}$ has a value of [1, 2] but then a transformation happens. $\hat{i}$ lands at [2, 0] and $\hat{j}$ lands at [0, 1.5]. Where does $\vec{v}$ land?

2. Vector $\vec{v}$ has a value of [1, 2] but then a transformation happens. $\hat{i}$ lands at [-2, 1] and $\hat{j}$ lands at [1, -2]. Where does $\vec{v}$ land?

3. A transformation $\hat{i}$ lands at [1, 0] and $\hat{j}$ lands at [2, 2]. What is the determinant of this transformation?

4. Can two or more linear transformations be done in single linear transformation? Why or why not?

5. Solve the system of equations for x, y, and z:

$$3x + 1y + 0z = = 54$$
$$2x + 4y + 1z = 12$$
$$3x + 1y + 8z = 6$$

6. Is the following matrix linearly dependent? Why or why not?

$$\begin{bmatrix} 2 & 1 \\ 6 & 3 \end{bmatrix}$$

Answers are in Appendix B.

## CHAPTER 5

Linear Regression

One of the most practical techniques in data analysis is fitting a line through observed data points to show a relationship between two or more variables. A *regression* attempts to fit a function to observed data to make predictions on new data. A *linear regression* fits a straight line to observed data, attempting to demonstrate a linear relationship between variables and make predictions on new data yet to be observed.

It might make more sense to see a picture rather than read a description of linear regression. There is an example of a linear regression in Figure 5-1.

Linear regression is a workhorse of data science and statistics and not only applies concepts we learned in previous chapters but sets up new foundations for later topics like neural networks (Chapter 7) and logistic regression (Chapter 6). This relatively simple technique has been around for more than two hundred years and contemporarily is branded as a form of machine learning.

Machine learning practitioners often take a different approach to validation, starting with a train-test split of the data. Statisticians are more likely to use metrics like prediction intervals and correlation for statistical significance. We will cover both schools of thought so readers can bridge the ever-widening gap between the two disciplines, and thus find themselves best equipped to wear both hats.

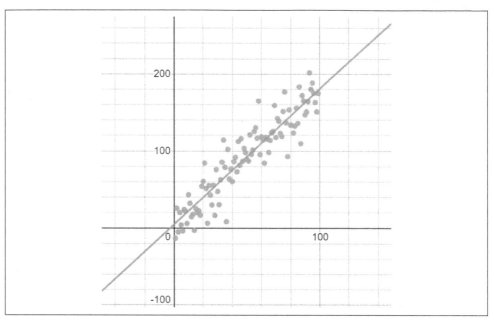

Figure 5-1. Example of a linear regression, which fits a line to observed data

Wait, Regression Is Machine Learning?

Machine learning has several techniques under its umbrella, but the one with the most use cases currently is supervised learning, and regressions play a big role here. This is why linear regression is branded as a form of machine learning. Confusingly, statisticians may refer to their regression models as *statistical learning*, whereas data science and machine learning professionals call their models *machine learning*.

While supervised learning is often regression, unsupervised machine learning is more about clustering and anomaly detection. Reinforcement learning often pairs supervised machine learning with simulation to rapidly generate synthetic data.

We will learn two more forms of supervised machine learning in Chapter 6 on logistic regression and Chapter 7 on neural networks.

A Basic Linear Regression

I want to study the relationship between the age of a dog and the number of veterinary visits it had. In a fabricated sample we have 10 random dogs. I am a fan of understanding complex techniques with simple datasets (real or otherwise), so we understand the strengths and limitations of the technique without complex data muddying the water. Let's plot this dataset as shown in Figure 5-2.

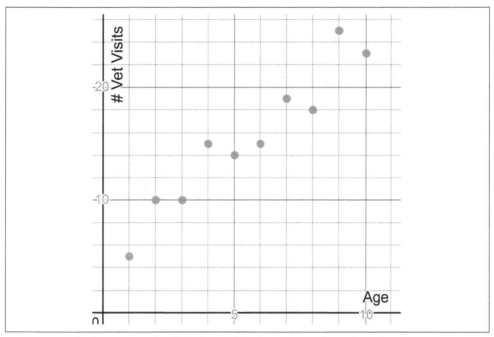

Figure 5-2. Plotting a sample of 10 dogs with their age and number of vet visits

We can clearly see there is a *linear correlation* here, meaning when one of these variables increases/decreases, the other increases/decreases in a roughly proportional amount. We could draw a line through these points to show a correlation like this in Figure 5-3.

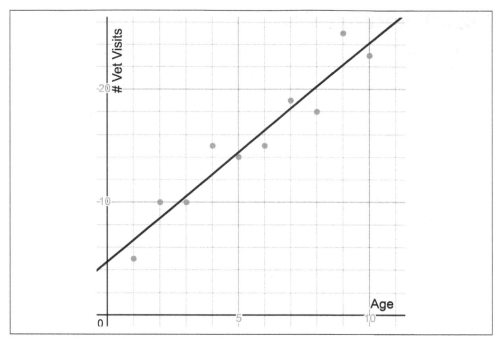

Figure 5-3. Fitting a line through our data

I will show how to calculate this fitted line later in this chapter. We will also explore how to calculate the quality of this fitted line. For now, let's focus on the benefits of performing a linear regression. It allows us to make predictions on data we have not seen before. I do not have a dog in my sample that is 8.5 years old, but I can look at this line and estimate the dog will have 21 veterinary visits in its life. I just look at the line where $x = 8.5$ and I see that $y = 21.218$ as shown in Figure 5-4. Another benefit is we can analyze variables for possible relationships and hypothesize that correlated variables are causal to one another.

Now what are the downsides of a linear regression? I cannot expect that every outcome is going to fall *exactly* on that line. After all, real-world data is noisy and never perfect and will not follow a straight line. It may not remotely follow a straight line at all! There is going to be error around that line, where the point will fall above or below the line. We will cover this mathematically when we talk about p-values, statistical significance, and prediction intervals, which describes how reliable our linear regression is. Another catch is we should not use the linear regression to make predictions outside the range of data we have, meaning we should not make predictions where $x < 0$ and $x > 10$ because we do not have data outside those values.

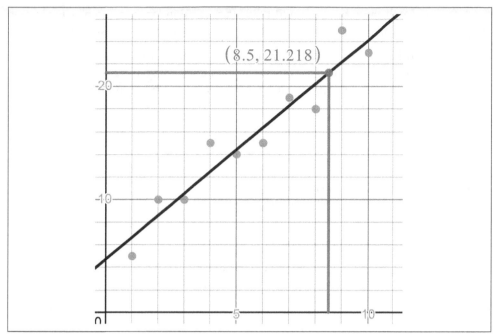

Figure 5-4. Making a prediction using a linear regression, seeing that an 8.5-year-old dog is predicted to have about 21.2 vet visits

Don't Forget Sampling Bias!

We should question this data and how it was sampled to detect bias. Was this at a single veterinary clinic? Multiple random clinics? Is there self-selection bias by using veterinary data, only polling dogs that visit the vet? If the dogs were sampled in the same geography, can that sway the data? Perhaps dogs in hot desert climates go to vets more for heat exhaustion and snake bites, and this would inflate our veterinary visits in our sample.

As discussed in Chapter 3, it has become fashionable to make data an oracle for truth. However data is simply a sample from a population, and we need to practice discernment on how well represented our sample is. Be just as interested (if not more) in where the data comes from and not just what the data says.

Basic Linear Regression with scikit-learn

We have a lot to learn regarding linear regression in this chapter, but let's start out with some code to execute what we know so far.

There are plenty of platforms to perform a linear regression, from Excel to Python and R. But we will stick with Python in this book, starting with scikit-learn to do the work for us. I will show how to build a linear regression "from scratch" later in this chapter so we grasp important concepts like gradient descent and least squares.

Example 5-1 is how we use scikit-learn to perform a basic, unvalidated linear regression on the sample of 10 dogs. We pull in this data using Pandas (*https://oreil.ly/ xCvwR*), convert it into NumPy arrays, perform linear regression using scikit-learn, and use Plotly to display it in a chart.

Example 5-1. Using scikit-learn to do a linear regression

```python
import pandas as pd
import matplotlib.pyplot as plt
from sklearn.linear_model import LinearRegression

# Import points
df = pd.read_csv('https://bit.ly/3goOAnt', delimiter=",")

# Extract input variables (all rows, all columns but last column)
X = df.values[:, :-1]

# Extract output column (all rows, last column)
Y = df.values[:, -1]

# Fit a line to the points
fit = LinearRegression().fit(X, Y)

# m = 1.7867224, b = -16.51923513
m = fit.coef_.flatten()
b = fit.intercept_.flatten()
print("m = {0}".format(m))
print("b = {0}".format(b))

# show in chart
plt.plot(X, Y, 'o') # scatterplot
plt.plot(X, m*X+b) # line
plt.show()
```

First we import the data from this CSV on GitHub (*https://bit.ly/3cIH97A*). We separate the two columns into X and Y datasets using Pandas. We then fit() the LinearRegression model to the input X data and the output Y data. We can then get the m and b coefficients that describe our fitted linear function.

In the plot, sure enough you will get a fitted line running through these points shown in Figure 5-5.

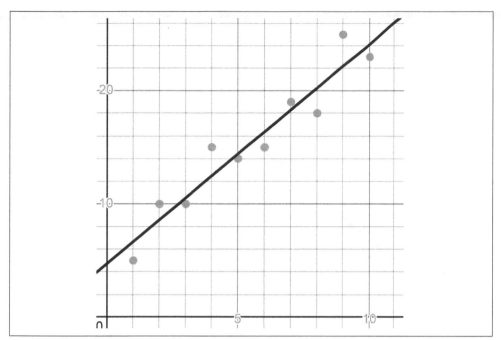

Figure 5-5. SciPy will fit a regression line to your data

What decides the best fit line to these points? Let's discuss that next.

Residuals and Squared Errors

How do statistics tools like scikit-learn come up with a line that fits to these points? It comes down to two questions that are fundamental to machine learning training:

- What defines a "best fit"?
- How do we get to that "best fit"?

The first question has a pretty established answer: we minimize the squares, or more specifically the sum of the squared residuals. Let's break that down. Draw any line through the points. The *residual* is the numeric difference between the line and the points, as shown in Figure 5-6.

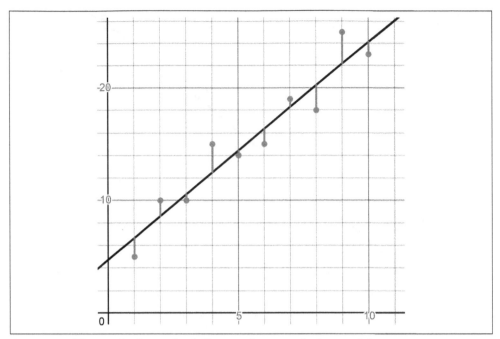

Figure 5-6. The residuals are the differences between the line and the points

Points above the line will have a positive residual, and points below the line will have a negative residual. In other words, it is the subtracted difference between the predicted y-values (derived from the line) and the actual y-values (which came from the data). Another name for residuals are *errors*, because they reflect how wrong our line is in predicting the data.

Let's calculate these differences between these 10 points and the line $y = 1.93939x + 4.73333$ in Example 5-2 and the residuals for each point in Example 5-3.

Example 5-2. Calculating the residuals for a given line and data

```
import pandas as pd

# Import points
points = pd.read_csv('https://bit.ly/3goOAnt', delimiter=",").itertuples()

# Test with a given line
m = 1.93939
b = 4.73333

# Calculate the residuals
for p in points:
    y_actual = p.y
    y_predict = m*p.x + b
```

```
residual = y_actual - y_predict
print(residual)
```

Example 5-3. The residuals for each point

```
-1.67272
1.3878900000000005
-0.5515000000000008
2.5091099999999997
-0.4302799999999998
-1.3696699999999993f
0.6909400000000012
-2.2484499999999983
2.812160000000002
-1.1272299999999973
```

If we are fitting a straight line through our 10 data points, we likely want to minimize these residuals in total so there is the least gap possible between the line and points. But how do we measure the "total"? The best approach is to take the *sum of squares*, which simply squares each residual, or multiplies each residual by itself, and sums them. We take each actual y-value and subtract from it the predicted y-value taken from the line, then square and sum all those differences.

Why Not Absolute Values?

You might wonder why we have to square the residuals before summing them. Why not just add them up without squaring? That will not work because the negatives will cancel out the positives. What if we add the absolute values, where we turn all negative values into positive values? That sounds promising, but absolute values are mathematically inconvenient. More specifically, absolute values do not work well with calculus derivatives that we are going to use later for gradient descent. This is why we choose the squared residuals as our way of totaling the loss.

A visual way to think of it is shown in Figure 5-7, where we overlay a square on each residual and each side is the length of the residual. We sum the area of all these squares, and later we will learn how to find the minimum sum we can achieve by identifying the best *m* and *b*.

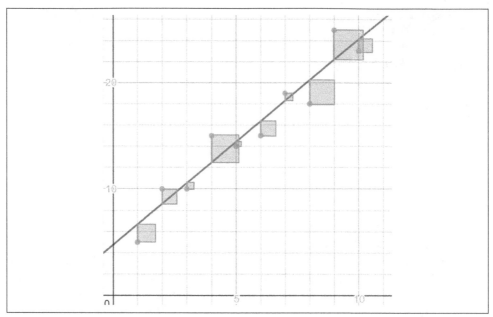

Figure 5-7. Visualizing the sum of squares, which would be the sum of all areas where each square has a side length equal to the residual

Let's modify our code in Example 5-4 to find the sum of squares.

Example 5-4. Calculating the sum of squares for a given line and data

```
import pandas as pd

# Import points
points = pd.read_csv("https://bit.ly/2KF29Bd").itertuples()

# Test with a given line
m = 1.93939
b = 4.73333

sum_of_squares = 0.0

# calculate sum of squares
for p in points:
    y_actual = p.y
    y_predict = m*p.x + b
    residual_squared = (y_predict - y_actual)**2
    sum_of_squares += residual_squared

print("sum of squares = {}".format(sum_of_squares))
# sum of squares = 28.096969704500005
```

Next question: how do we find the m and b values that will produce the minimum sum of squares, without using a library like scikit-learn? Let's look at that next.

Finding the Best Fit Line

We now have a way to measure the quality of a given line against the data points: the sum of squares. The lower we can make that number, the better the fit. Now how do we find the right m and b values that create the *least* sum of squares?

There are a couple of search algorithms we can employ, which try to find the right set of values to solve a given problem. You can try a *brute force* approach, generating random m and b values millions of times and choosing the ones that produce the least sum of squares. This will not work well because it will take an endless amount of time to find even a decent approximation. We will need something a little more guided. I will curate five techniques you can use: closed form, matrix inversion, matrix decomposition, gradient descent, and stochastic gradient descent. There are other search algorithms like hill climbing that could be used (and are covered in Appendix A), but we will stick with what's common.

Machine Learning Training Is Fitting a Regression

This is the heart of "training" a machine learning algorithm. We provide some data and an objective function (the sum of squares) and it will find the right coefficients m and b to fulfill that objective. So when we "train" a machine learning model we really are minimizing a loss function.

Closed Form Equation

Some readers may ask if there is a formula (called a *closed form equation*) to fit a linear regression by exact calculation. The answer is yes, but only for a simple linear regression with one input variable. This luxury does not exist for many machine learning problems with several input variables and a large amount of data. We can use linear algebra techniques to scale up, and we will talk about this shortly. We will also take the opportunity to learn about search algorithms like stochastic gradient descent.

For a simple linear regression with only one input and one output variable, here are the closed form equations to calculate m and b. Example 5-5 shows how you can do these calculations in Python.

$$m = \frac{n\Sigma xy - \Sigma x \Sigma y}{n\Sigma x^2 - (\Sigma x)^2}$$

$$b = \frac{\sum y}{n} - m\frac{\sum x}{n}$$

Example 5-5. Calculating m and b for a simple linear regression

```
import pandas as pd

# Load the data
points = list(pd.read_csv('https://bit.ly/2KF29Bd', delimiter=",").itertuples())

n = len(points)

m = (n*sum(p.x*p.y for p in points) - sum(p.x for p in points) *
    sum(p.y for p in points)) / (n*sum(p.x**2 for p in points) -
    sum(p.x for p in points)**2)

b = (sum(p.y for p in points) / n) - m * sum(p.x for p in points) / n

print(m, b)
# 1.9393939393939394 4.7333333333333325
```

These equations to calculate *m* and *b* are derived from calculus, and we will do some calculus work with SymPy later in this chapter if you have the itch to discover where formulas come from. For now, you can plug in the number of data points *n* as well as iterate the *x*- and *y*-values to do the operations just described.

Going forward, we will learn approaches that are more oriented to contemporary techniques that cope with larger amounts of data. Closed form equations tend not to scale well.

Computational Complexity

The reason the closed form equations do not scale well with larger datasets is due to a computer science concept called *computational complexity*, which measures how long an algorithm takes as a problem size grows. This might be worth getting familiar with; here are two great YouTube videos on the topic:

- "P vs. NP and the Computational Complexity Zoo" (*https://oreil.ly/TzQBl*)
- "What Is Big O Notation?" (*https://oreil.ly/EjcSR*)

Inverse Matrix Techniques

Going forward, I will sometimes alternate the coefficients *m* and *b* with different names, β_1 and β_0, respectively. This is the convention you will see more often in the professional world, so it might be a good time to graduate.

While we dedicated an entire chapter to linear algebra in Chapter 4, applying it can be a bit overwhelming when you are new to math and data science. This is why most examples in this book will use plain Python or scikit-learn. However, I will sprinkle in linear algebra where it makes sense, just to show how linear algebra is useful. If you find this section overwhelming, feel free to move on to the rest of the chapter and come back later.

We can use transposed and inverse matrices, which we covered in Chapter 4, to fit a linear regression. Next, we calculate a vector of coefficients b given a matrix of input variable values X and a vector of output variable values y. Without going down a rabbit hole of calculus and linear algebra proofs, here is the formula:

$$b = \left(X^T \cdot X \right)^{-1} \cdot X^T \cdot y$$

You will notice transposed and inverse operations are performed on the matrix X and combined with matrix multiplication. Here is how we perform this operation in NumPy in Example 5-6 to get our coefficients m and b.

Example 5-6. Using inverse and transposed matrices to fit a linear regression

```
import pandas as pd
from numpy.linalg import inv
import numpy as np

# Import points
df = pd.read_csv('https://bit.ly/3goOAnt', delimiter=",")

# Extract input variables (all rows, all columns but last column)
X = df.values[:, :-1].flatten()

# Add placeholder "1" column to generate intercept
X_1 = np.vstack([X, np.ones(len(X))]).T

# Extract output column (all rows, last column)
Y = df.values[:, -1]

# Calculate coefficents for slope and intercept
b = inv(X_1.transpose() @ X_1) @ (X_1.transpose() @ Y)
print(b) # [1.93939394, 4.73333333]

# Predict against the y-values
y_predict = X_1.dot(b)
```

It is not intuitive, but note we have to stack a "column" of 1s next to our X column. The reason is this will generate the intercept β_0 coefficient. Since this column is all 1s, it effectively generates the intercept and not just a slope β_1.

Matrix Decomposition

When you have a lot of data with a lot of dimensions, computers can start to choke and produce unstable results. This is a use case for matrix decomposition, which we learned about in Chapter 4 on linear algebra. In this specific case, we take our matrix X, append an additional column of 1s to generate the intercept β_0 just like before, and then decompose it into two component matrices Q and R:

$$X = Q \cdot R$$

Avoiding more calculus rabbit holes, here is how we use Q and R to find the beta coefficient values in the matrix form b:

$$b = R^{-1} \cdot Q^T \cdot y$$

And Example 5-7 shows how we use the preceding QR decomposition formula in Python using NumPy to perform a linear regression.

Example 5-7. Using QR decomposition to perform a linear regression

```python
import pandas as pd
from numpy.linalg import qr, inv
import numpy as np

# Import points
df = pd.read_csv('https://bit.ly/3goOAnt', delimiter=",")

# Extract input variables (all rows, all columns but last column)
X = df.values[:, :-1].flatten()

# Add placeholder "1" column to generate intercept
X_1 = np.vstack([X, np.ones(len(X))]).transpose()

# Extract output column (all rows, last column)
Y = df.values[:, -1]

# calculate coefficents for slope and intercept
# using QR decomposition
Q, R = qr(X_1)
b = inv(R).dot(Q.transpose()).dot(Y)

print(b) # [1.93939394, 4.73333333]
```

Typically, *QR* decomposition is the method used by many scientific libraries for linear regression because it copes with large amounts of data more easily and is more stable. What do I mean by *stable*? *Numerical stability (https://oreil.ly/A4BWJ)*

is how well an algorithm keeps errors minimized, rather than amplifying errors in approximations. Remember that computers work only to so many decimal places and have to approximate, so it becomes important our algorithms do not deteriorate with compounding errors in those approximations.

Overwhelmed?

If you find these linear algebra examples of linear regression overwhelming, do not worry! I just wanted to provide exposure to a practical use case for linear algebra. Going forward, we will focus on other techniques you can use.

Gradient Descent

Gradient descent is an optimization technique that uses derivatives and iterations to minimize/maximize a set of parameters against an objective. To learn about gradient descent, let's do a quick thought experiment and then apply it on a simple example.

A Thought Experiment on Gradient Descent

Imagine you are in a mountain range at night and given a flashlight. You are trying to get to the lowest point of the mountain range. You can see the slope around you before you even take a step. You step in directions where the slope visibly goes downward. You take bigger steps for bigger slopes, and smaller steps for smaller slopes. Ultimately, you will find yourself at a low point where the slope is flat, a value of 0. Sounds pretty good, right? This approach with the flashlight is known as *gradient descent*, where we step in directions where the slope goes downward.

In machine learning, we often think of all possible sum of square losses we will encounter with different parameters as a mountainous landscape. We want to minimize our loss, and we navigate the loss landscape to do it. To solve this problem, gradient descent has an attractive feature: the partial derivative is that flashlight, allowing us to see the slopes for every parameter (in this case m and b, or β_0 and β_1). We step in directions for m and b where the slope goes downward. We take bigger steps for bigger slopes and smaller steps for smaller slopes. We can simply calculate the length of this step by taking a fraction of the slope. This fraction is known as our *learning rate*. The higher the learning rate, the faster it will run at the cost of accuracy. But the lower the learning rate, the longer it will take to train and require more iterations.

Deciding a learning rate is like choosing between an ant, a human, or a giant to step down the slope. An ant (small learning rate) will take tiny steps and take an unacceptably long time to get to the bottom but will do so precisely. A giant (large learning rate) may keep stepping over the minimum to the point he may never reach it no matter how many steps he takes. The human (moderate learning rate) probably has most balanced step size, having the right trade between speed and accuracy in arriving at the minimum.

Let's Walk Before We Run

For the function $f(x) = (x - 3)^2 + 4$, let's find the x-value that produces the lowest point of that function. While we could solve this algebraically, let's use gradient descent to do it.

Here is visually what we are trying to do. As shown in Figure 5-8, we want to "step" x toward the minimum where the slope is 0.

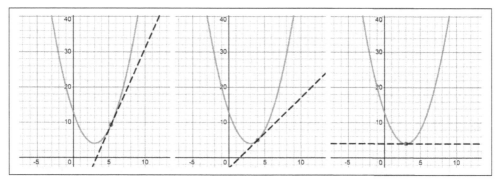

Figure 5-8. Stepping toward the local minimum where the slope approaches 0

In Example 5-8, the function f(x) and its derivative with respect to x is dx_f(x). Recall we covered in Chapter 1 how to use SymPy to calculate derivatives. After finding the derivative, we then proceed to perform gradient descent.

Example 5-8. Using gradient descent to find the minimum of a parabola

```
import random

def f(x):
    return (x - 3) ** 2 + 4

def dx_f(x):
    return 2*(x - 3)

# The learning rate
L = 0.001

# The number of iterations to perform gradient descent
iterations = 100_000

 # start at a random x
x = random.randint(-15,15)
```

```
for i in range(iterations):

    # get slope
    d_x = dx_f(x)

    # update x by subtracting the (learning rate) * (slope)
    x -= L * d_x

print(x, f(x)) # prints 2.999999999999889 4.0
```

If we graph the function (as shown in Figure 5-8), we should see the lowest point of the function is clearly where $x = 3$, and the preceding code should get very close to that. The learning rate is used to take a fraction of the slope and subtract it from the x-value on each iteration. Bigger slopes will result in bigger steps, and smaller slopes will result in smaller steps. After enough iterations, x will end up at the lowest point of the function (or close enough to it) where the slope is 0.

Gradient Descent and Linear Regression

You now might be wondering how we use this for linear regression. Well, it's the same idea except our "variables" are m and b (or β_0 and β_1) rather than x. Here's why: in a simple linear regression we already know the x- and y-values because those are provided as the training data. The "variables" we need to solve are actually the parameters m and b, so we can find the best fit line that will then accept an x variable to predict a new y-value.

How do we calculate the slopes for m and b? We need the partial derivatives for each of these. What function are we taking the derivative of? Remember we are trying to minimize loss and that will be the sum of squares. So we need to find the derivatives of our sum of squares function with respect to m and b.

I implement these two partial derivatives for m and b as shown in Example 5-9. We will learn how to do this shortly in SymPy. I then perform gradient descent to find m and b: 100,000 iterations with a learning rate of .001 will be sufficient. Note that the smaller you make that learning rate, the slower it will be and the more iterations you will need. But if you make it too high, it will run fast but have a poor approximation. When someone says a machine learning algorithm is "learning" or "training," it really is just fitting a regression like this.

Example 5-9. Performing gradient descent for a linear regression

```
import pandas as pd

# Import points from CSV
points = list(pd.read_csv("https://bit.ly/2KF29Bd").itertuples())

# Building the model
```

```
m = 0.0
b = 0.0


# The learning Rate
L = .001

# The number of iterations
iterations = 100_000

n = float(len(points))  # Number of elements in X

# Perform Gradient Descent
for i in range(iterations):

    # slope with respect to m
    D_m = sum(2 * p.x * ((m * p.x + b) - p.y) for p in points)

    # slope with respect to b
    D_b = sum(2 * ((m * p.x + b) - p.y) for p in points)

    # update m and b
    m -= L * D_m
    b -= L * D_b

print("y = {0}x + {1}".format(m, b))
# y = 1.9393939393939548x + 4.733333333333227
```

Well, not bad! That approximation got close to our closed form equation solution.
But what's the catch? Just because we found the "best fit line" by minimizing sum of
squares, that does not mean our linear regression is any good. Does minimizing the
sum of squares guarantee a great model to make predictions? Not exactly. Now that I
showed you how to fit a linear regression, let's take a step back, revisit the big picture,
and determine whether a given linear regression is the right way to make predictions
in the first place. But before we do that, here's one more detour showing the SymPy
solution.

Gradient Descent for Linear Regression Using SymPy

If you want the SymPy code that came up with these two derivatives for the sum of
squares function, for *m* and *b* respectively, here is the code in Example 5-10.

Example 5-10. Calculating partial derivatives for m and b

```
from sympy import *

m, b, i, n = symbols('m b i n')
x, y = symbols('x y', cls=Function)
```

```
sum_of_squares = Sum((m*x(i) + b - y(i)) ** 2, (i, 0, n))

d_m = diff(sum_of_squares, m)
d_b = diff(sum_of_squares, b)
print(d_m)
print(d_b)

# OUTPUTS
# Sum(2*(b + m*x(i) - y(i))*x(i), (i, 0, n))
# Sum(2*b + 2*m*x(i) - 2*y(i), (i, 0, n))
```

You will see the two derivatives for *m* and *b*, respectively, printed. Note the Sum() function will iterate and add items together (in this case all the data points), and we treat *x* and *y* as functions that look up a value for a given point at index *i*.

In mathematical notation, where *e(x)* represents the sum of squares loss function, here are the partial derivatives for *m* and *b*:

$$e(x) = \sum_{i=0}^{n} \left((mx_i + b) - y_i \right)^2$$

$$\frac{d}{dm} e(x) = \sum_{i=0}^{n} 2(b + mx_i - y_i)x_i$$

$$\frac{d}{db} e(x) = \sum_{i=0}^{n} (2b + 2mx_i - 2y_i)$$

If you want to apply our dataset and execute a linear regression using gradient descent, you will have to perform a few additional steps as shown in Example 5-11. We will need to substitute for the n, x(i) and y(i) values, iterating all of our data points for the d_m and d_b derivative functions. That should leave only the m and b variables, which we will search for the optimal values using gradient descent.

Example 5-11. Solving linear regression using SymP

```
import pandas as pd
from sympy import *

# Import points from CSV
points = list(pd.read_csv("https://bit.ly/2KF29Bd").itertuples())

m, b, i, n = symbols('m b i n')
x, y = symbols('x y', cls=Function)

sum_of_squares = Sum((m*x(i) + b - y(i)) ** 2, (i, 0, n))

d_m = diff(sum_of_squares, m) \
    .subs(n, len(points) - 1).doit() \
```

```
        .replace(x, lambda i: points[i].x) \
        .replace(y, lambda i: points[i].y)

d_b = diff(sum_of_squares, b) \
    .subs(n, len(points) - 1).doit() \
    .replace(x, lambda i: points[i].x) \
    .replace(y, lambda i: points[i].y)

# compile using lambdify for faster computation
d_m = lambdify([m, b], d_m)
d_b = lambdify([m, b], d_b)

# Building the model
m = 0.0
b = 0.0

# The learning Rate
L = .001

# The number of iterations
iterations = 100_000

# Perform Gradient Descent
for i in range(iterations):

    # update m and b
    m -= d_m(m,b) * L
    b -= d_b(m,b) * L

print("y = {0}x + {1}".format(m, b))
# y = 1.939393939393954x + 4.733333333333231
```

As shown in Example 5-11, it is a good idea to call `lambdify()` on both of our partial derivative functions to convert them from SymPy to an optimized Python function. This will cause computations to perform much more quickly when we do gradient descent. The resulting Python functions are backed by NumPy, SciPy, or whatever numerical libraries SymPy detects you have available. After that, we can perform gradient descent.

Finally, if you are curious about what the loss function looks like for this simple linear regression, Example 5-12 shows the SymPy code that plugs the x, y, and n values into our loss function and then plots m and b as the input variables. Our gradient descent algorithm gets us to the lowest point in this loss landscape shown in Figure 5-9.

Example 5-12. Plotting the loss function for linear regression

```
from sympy import *
from sympy.plotting import plot3d
import pandas as pd
```

```
points = list(pd.read_csv("https://bit.ly/2KF29Bd").itertuples())
m, b, i, n = symbols('m b i n')
x, y = symbols('x y', cls=Function)

sum_of_squares = Sum((m*x(i) + b - y(i)) ** 2, (i, 0, n)) \
    .subs(n, len(points) - 1).doit() \
    .replace(x, lambda i: points[i].x) \
    .replace(y, lambda i: points[i].y)

plot3d(sum_of_squares)
```

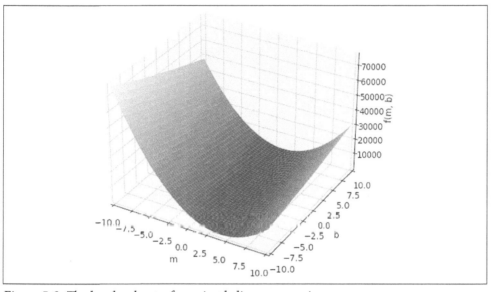

Figure 5-9. The loss landscape for a simple linear regression

Overfitting and Variance

Riddle me this: if we truly wanted to minimize loss, as in reduce the sum of squares to 0, what would we do? Are there options other than linear regression? One conclusion you may arrive at is simply fit a curve that touches all the points. Heck, why not just connect the points in segments and use that to make predictions as shown in Figure 5-10? That gives us a loss of 0!

Shoot, why did we go through all that trouble with linear regression and not do this instead? Well, remember our big-picture objective is not to minimize the sum of squares but to make accurate predictions on new data. This connect-the-dots model is severely *overfit*, meaning it shaped the regression to the training data too exactly to the point it will predict poorly on new data. This simple connect-the-dots model is sensitive to outliers that are far away from the rest of the points, meaning it will have high *variance* in predictions. While the points in this example are relatively close to a line, this problem will be a lot worse with other datasets with more spread and outliers. Because overfitting increases variance, predictions are going to be all over the place!

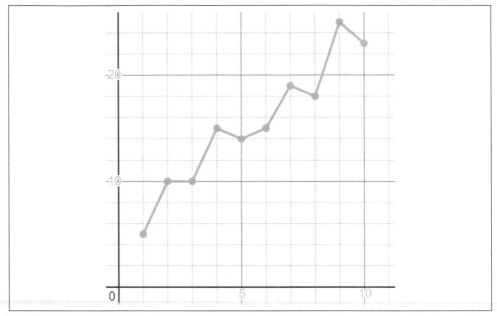

Figure 5-10. Performing a regression by simply connecting the points, resulting in zero loss

Overfitting Is Memorization

When you hear someone say a regression "memorized" the data rather than generalizing it, they are talking about overfitting.

As you can guess, we want to find effective generalizations in our model rather than memorize data. Otherwise, our regression simply turns into a database where we look up values.

This is why in machine learning you will find bias is added to the model, and linear regression is considered a highly biased model. This is not the same as bias in the data, which we talked about extensively in Chapter 3. *Bias in a model* means we prioritize a method (e.g., maintaining a straight line) as opposed to bending and fitting to exactly what the data says. A biased model leaves some wiggle room hoping to minimize loss on new data for better predictions, as opposed to minimizing loss on data it was trained on. I guess you could say adding bias to a model counteracts overfitting with *underfitting*, or fitting less to the training data.

As you can imagine, this is a balancing act because it is two contradictory objectives. In machine learning, we basically are saying, "I want to fit a regression to my data, but I don't want to fit it *too much*. I need some wiggle room for predictions on new data that will be different."

Lasso and Ridge Regression

Two somewhat popular variants of linear regression are lasso regression and ridge regression. Ridge regression adds a further bias to a linear regression in the form of a penalty, therefore causing it to fit less to the data. Lasso regression will attempt to marginalize noisy variables, making it useful when you want to automatically remove variables that might be irrelevant.

Still, we cannot just apply a linear regression to some data, make some predictions with it, and assume all is OK. A linear regression can overfit even with the bias of a straight line. Therefore, we need to check and mitigate for both overfitting and underfitting to find the sweet spot between the two. That is, unless there is not one at all, in which case you should abandon the model altogether.

Stochastic Gradient Descent

In a machine learning context, you are unlikely to do gradient descent in practice like we did earlier, where we trained on all training data (called *batch gradient descent*). In practice, you are more likely to perform *stochastic gradient descent*, which will train on only one sample of the dataset on each iteration. In *mini-batch gradient descent*, multiple samples of the dataset are used (e.g., 10 or 100 data points) on each iteration.

Why use only part of the data on each iteration? Machine learning practitioners cite a few benefits. First, it reduces computation significantly, as each iteration does not have to traverse the entire training dataset but only part of it. The second benefit is it reduces overfitting. Exposing the training algorithm to only part of the data on each iteration keeps changing the loss landscape so it does not settle in the loss minimum. After all, minimizing the loss is what causes overfitting and so we introduce some randomness to create a little bit of underfitting (but hopefully not too much).

Of course, our approximation becomes loose so we have to be careful. This is why we will talk about train/test splits shortly, as well as other metrics to evaluate our linear regression's reliability.

Example 5-13 shows how to perform stochastic gradient descent in Python. If you change the sample size to be more than 1, it will perform mini-batch gradient descent.

Example 5-13. Performing stochastic gradient descent for a linear regression

```python
import pandas as pd
import numpy as np

# Input data
data = pd.read_csv('https://bit.ly/2KF29Bd', header=0)

X = data.iloc[:, 0].values
Y = data.iloc[:, 1].values

n = data.shape[0]  # rows

# Building the model
m = 0.0
b = 0.0

sample_size = 1  # sample size
L = .0001  # The learning Rate
epochs = 1_000_000  # The number of iterations to perform gradient descent

# Performing Stochastic Gradient Descent
for i in range(epochs):
    idx = np.random.choice(n, sample_size, replace=False)
    x_sample = X[idx]
    y_sample = Y[idx]

    # The current predicted value of Y
    Y_pred = m * x_sample + b

    # d/dm derivative of loss function
    D_m = (-2 / sample_size) * sum(x_sample * (y_sample - Y_pred))

    # d/db derivative of loss function
    D_b = (-2 / sample_size) * sum(y_sample - Y_pred)
    m = m - L * D_m  # Update m
    b = b - L * D_b  # Update b

    # print progress
    if i % 10000 == 0:
        print(i, m, b)

print("y = {0}x + {1}".format(m, b))
```

When I ran this, I got a linear regression of $y = 1.9382830354181135x + 4.753408787648379$. Obviously, your results are going to be different, and because of stochastic gradient descent we really aren't going to converge toward a specific minimum but will end up in a broader neighborhood.

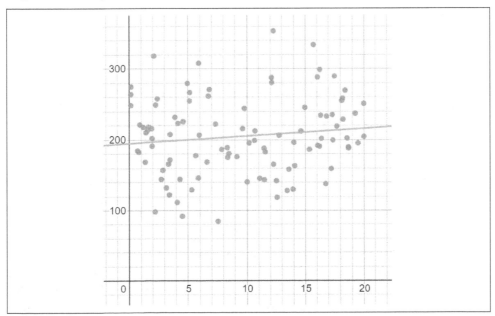

Is Randomness Bad?

If this randomness feels uncomfortable where you get a different answer every time you run a piece of code, welcome to the world of machine learning, optimization, and stochastic algorithms! Many algorithms that do approximations are random-based, and while some are extremely useful, others can be sloppy and perform poorly, as you might expect.

A lot of people look to machine learning and AI as some tool that gives objective and precise answers, but that cannot be farther from the truth. Machine learning produces approximations with a degree of uncertainty, often without ground truth once in production. Machine learning can be misused if one is not aware of how it works, and it is remiss not to acknowledge its nondeterministic and approximate nature.

While randomness can create some powerful tools, it can also be abused. Be careful to not use seed values and randomness to p-hack a "good" result, and put effort into analyzing your data and model.

The Correlation Coefficient

Take a look at this scatterplot in Figure 5-11 alongside its linear regression. Why might a linear regression not work too well here?

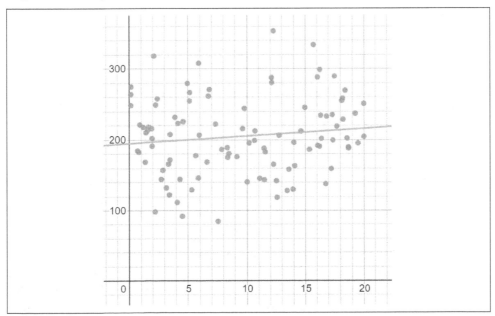

Figure 5-11. A scatterplot of data with high variance

The problem here is that the data has high variance. If the data is extremely spread out, it is going to drive up the variance to the point predictions become less accurate and useful, resulting in large residuals. Of course we can introduce a more biased model, such as linear regression, to not bend and respond to the variance so easily. However, the underfitting is also going to undermine our predictions because the data is so spread out. We need to numerically measure how "off" our predictions are.

So how do you measure these residuals in aggregate? How do you also get a sense for how bad the variance in the data is? Let me introduce you to the *correlation coefficient*, also called the *Pearson correlation*, which measures the strength of the relationship between two variables as a value between –1 and 1. A correlation coefficient closer to 0 indicates there is no correlation. A correlation coefficient closer to 1 indicates a strong *positive correlation*, meaning when one variable increases, the other proportionally increases. If it is closer to –1 then it indicates a strong *negative correlation*, which means as one variable increases the other proportionally decreases.

Note the correlation coefficient is often denoted as r. The highly scattered data in Figure 5-11 has a correlation coefficient of 0.1201. Since it is much closer to 0 than 1, we can infer the data has little correlation.

Here are four other scatterplots in Figure 5-12 showing their correlation coefficients. Notice that the more the points follow a line, the stronger the correlation. More dispersed points result in weaker correlations.

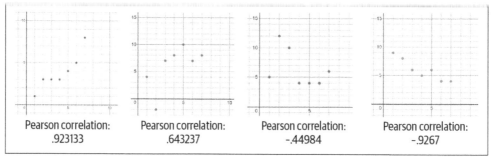

| Pearson correlation: .923133 | Pearson correlation: .643237 | Pearson correlation: -.44984 | Pearson correlation: -.9267 |

Figure 5-12. Correlation coefficients for four scatterplots

As you can imagine, the correlation coefficient is useful for seeing if there is a possible relationship between two variables. If there is a strong positive-negative relationship, it will be useful in our linear regression. If there is not a relationship, they may just add noise and hurt model accuracy.

How do we use Python to calculate the correlation coefficient? Let's use the simple 10-point dataset (*https://bit.ly/2KF29Bd*) we used earlier. A quick and easy way to analyze correlations for all pairs of variables is using Pandas's corr() function. This makes it easy to see the correlation coefficient between every pair of variables in a dataset, which in this case will only be x and y. This is known as a *correlation matrix*. Take a look in Example 5-14.

Example 5-14. Using Pandas to see the correlation coefficient between every pair of variables

```
import pandas as pd

# Read data into Pandas dataframe
df = pd.read_csv('https://bit.ly/2KF29Bd', delimiter=",")

# Print correlations between variables
correlations = df.corr(method='pearson')
print(correlations)

# OUTPUT:
#            x         y
# x   1.000000  0.957586
# y   0.957586  1.000000
```

As you can see, the correlation coefficient 0.957586 between x and y indicates a strong positive correlation between the two variables. You can ignore the parts of the matrix where x or y is set to itself and has a value of 1.0. Obviously, when x or y is set to itself, the correlation will be perfect at 1.0, because the values match themselves exactly. When you have more than two variables, the correlation matrix will show a larger grid because there are more variables to pair and compare.

If you change the code to use a different dataset with a lot of variance, where the data is spread out, you will see that the correlation coefficient decreases. This again indicates a weaker correlation.

Calculating the Correlation Coefficient

For those who are mathematically curious how the correlation coefficient is calculated, here is the formula:

$$r = \frac{n\sum xy - (\sum x)(\sum y)}{\sqrt{n\sum x^2 - (\sum x^2)}\sqrt{n\sum y^2 - (\sum y^2)}}$$

If you want to see this fully implemented in Python, I like to use one-line for loops to do those summations. Example 5-15 shows a correlation coefficient implemented from scratch in Python.

Example 5-15. Calculating correlation coefficient from scratch in Python

```python
import pandas as pd
from math import sqrt

# Import points from CSV
points = list(pd.read_csv("https://bit.ly/2KF29Bd").itertuples())
n = len(points)

numerator = n * sum(p.x * p.y for p in points) - \
            sum(p.x for p in points) * sum(p.y for p in points)

denominator = sqrt(n*sum(p.x**2 for p in points) - sum(p.x for p in points)**2) \
                * sqrt(n*sum(p.y**2 for p in points) - sum(p.y for p in points)**2)

corr = numerator / denominator

print(corr)

# OUTPUT:
# 0.9575860952087218
```

Statistical Significance

Here is another aspect to a linear regression you must consider: is my data correlation coincidental? In Chapter 3 we studied hypothesis testing and p-values, and we are going to extend those ideas here with a linear regression.

The statsmodel Library

While we are not going to introduce yet another library in this book, it is worth mentioning that statsmodel (*https://oreil.ly/8oEHo*) is noteworthy if you want to do statistical analysis.

Scikit-learn and other machine learning libraries do not provide tools for statistical significance and confidence intervals for reasons we will discuss in another sidebar. We are going to code those techniques ourselves. But know this library exists and it is worth checking out!

Let's start with a fundamental question: is it possible I see a linear relationship in my data due to random chance? How can we be 95% sure the correlation between these two variables is significant and not coincidental? If this sounds like a hypothesis test from Chapter 3, it's because it is! We need to not just express the correlation coefficient but also quantify how confident we are that the correlation coefficient did not occur by chance.

Rather than estimating a mean like we did in Chapter 3 with the drug-testing example, we are estimating the population correlation coefficient based on a sample. We denote the population correlation coefficient with the Greek symbol ρ (Rho) while our sample correlation coefficient is r. Just like we did in Chapter 3, we will have a null hypothesis H_0 and alternative hypothesis H_1:

$H_0: \rho = 0$ (implies no relationship)
$H_1: \rho \neq 0$ (relationship is present)

Our null hypothesis H_0 is that there is no relationship between two variables, or more technically, the correlation coefficient is 0. The alternative hypothesis H_1 is there is a relationship, and it can be a positive or negative correlation. This is why the alternative hypothesis is defined as $\rho \neq 0$ to support both a positive and negative correlation.

Let's return to our dataset of 10 points as shown in Figure 5-13. How likely is it we would see these data points by chance? And they happen to produce what looks like a linear relationship?

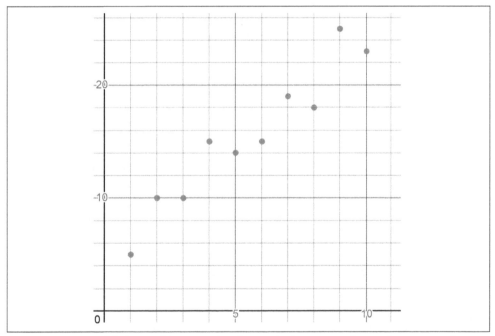

Figure 5-13. How likely would we see this data, which seems to have a linear correlation, by random chance?

We already calculated the correlation coefficient for this dataset in Example 5-14, which is 0.957586. That's a strong and compelling positive correlation. But again, we need to evaluate if this was by random luck. Let's pursue our hypothesis test with 95% confidence using a two-tailed test, exploring if there is a relationship between these two variables.

We talked about the T-distribution in Chapter 3, which has fatter tails to capture more variance and uncertainty. We use a T-distribution rather than a normal distribution to do hypothesis testing with linear regression. First, let's plot a T-distribution with a 95% critical value range as shown in Figure 5-14. We account for the fact there are 10 records in our sample and therefore we have 9 degrees of freedom (10 − 1 = 9).

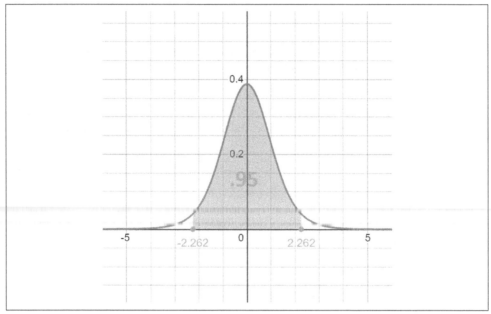

Figure 5-14. A T-distribution with 9 degrees of freedom, as there are 10 records and we subtract 1

The critical value is approximately ±2.262, and we can calculate that in Python as shown in Example 5-16. This captures 95% of the center area of our T-distribution.

Example 5-16. Calculating the critical value from a T-distribution

```
from scipy.stats import t

n = 10
lower_cv = t(n-1).ppf(.025)
upper_cv = t(n-1).ppf(.975)

print(lower_cv, upper_cv)
# -2.262157162740992 2.2621571627409915
```

If our test value happens to fall outside this range of (−2.262, 2.262), then we can reject our null hypothesis. To calculate the test value t, we need to use the following formula. Again r is the correlation coefficient and n is the sample size:

$$t = \frac{r}{\sqrt{\frac{1 - r^2}{n - 2}}}$$

$$t = \frac{.957586}{\sqrt{\frac{1 - .957586^2}{10 - 2}}} = 9.339956$$

Let's put the whole test together in Python as shown in Example 5-17. If our test value falls outside the critical range of 95% confidence, we accept that our correlation was not by chance.

Example 5-17. Testing significance for linear-looking data

```
from scipy.stats import t
from math import sqrt

# sample size
n = 10

lower_cv = t(n-1).ppf(.025)
upper_cv = t(n-1).ppf(.975)

# correlation coefficient
# derived from data https://bit.ly/2KF29Bd
r = 0.957586

# Perform the test
test_value = r / sqrt((1-r**2) / (n-2))

print("TEST VALUE: {}".format(test_value))
print("CRITICAL RANGE: {}, {}".format(lower_cv, upper_cv))

if test_value < lower_cv or test_value > upper_cv:
    print("CORRELATION PROVEN, REJECT H0")
else:
    print("CORRELATION NOT PROVEN, FAILED TO REJECT H0 ")

# Calculate p-value
if test_value > 0:
    p_value = 1.0 - t(n-1).cdf(test_value)
else:
    p_value = t(n-1).cdf(test_value)
```

```
# Two-tailed, so multiply by 2
p_value = p_value * 2
print("P-VALUE: {}".format(p_value))
```

The test value here is approximately 9.39956, which is definitely outside the range of (–2.262, 2.262) so we can reject the null hypothesis and say our correlation is real. That's because the p-value is remarkably significant: .000005976. This is well below our .05 threshold, so this is virtually not coincidence: there is a correlation. It makes sense the p-value is so small because the points strongly resemble a line. It is highly unlikely these points randomly arranged themselves near a line this closely by chance.

Figure 5-15 shows some other datasets with their correlation coefficients and p-values. Analyze each one of them. Which one is likely the most useful for predictions? What are the problems with the other ones?

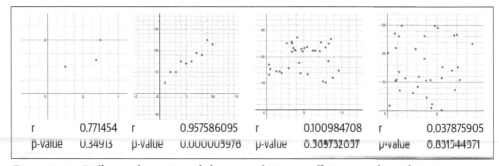

| r | 0.771454 | r | 0.957586095 | r | 0.100984708 | r | 0.037875905 |
| p-value | 0.34913 | p-value | 0.000005976 | p-value | 0.309732037 | p-value | 0.831544971 |

Figure 5-15. Different datasets and their correlation coefficients and p-values

Now that you've had a chance to work through the analysis on the datasets from Figure 5-15, let's walk through the findings. The left figure has a high positive correlation, but it only has three points. The lack of data drives up the p-value significantly to 0.34913 and increases the likelihood the data happened by chance. This makes sense because having just three data points makes it likely to see a linear pattern, but it's not much better than having two points that will simply connect a line between them. This brings up an important rule: having more data will decrease your p-value, especially if that data gravitates toward a line.

The second figure is what we just covered. It has only 10 data points, but it forms a linear pattern so nicely, we not only have a strong positive correlation but also an extremely low p-value. When you have a p-value this low, you can bet you are measuring an engineered and tightly controlled process, not something sociological or natural.

The right two images in Figure 5-15 fail to identify a linear relationship. Their correlation coefficient is close to 0, indicating no correlation, and the p-values unsurprisingly indicate randomness played a role.

The rule is this: the more data you have that consistently resembles a line, the more significant the p-value for your correlation will be. The more scattered or sparse the data, the more the p-value will increase and thus indicate your correlation occurred by random chance.

Coefficient of Determination

Let's learn an important metric you will see a lot in statistics and machine learning regressions. The *coefficient of determination*, called r^2, measures how much variation in one variable is explainable by the variation of the other variable. It is also the square of the correlation coefficient r. As r approaches a perfect correlation (–1 or 1), r^2 approaches 1. Essentially, r^2 shows how much two variables interact with each other.

Let's continue looking at our data from Figure 5-13. In Example 5-18, take our dataframe code from earlier that calculates the correlation coefficient and then simply square it. That will multiply each correlation coefficient by itself.

Example 5-18. Creating a correlation matrix in Pandas

```
import pandas as pd

# Read data into Pandas dataframe
df = pd.read_csv('https://bit.ly/2KF29Bd', delimiter=",")

# Print correlations between variables
coeff_determination = df.corr(method='pearson') ** 2
print(coeff_determination)

# OUTPUT:
#           x         y
# x  1.000000  0.916971
# y  0.916971  1.000000
```

A coefficient of determination of 0.916971 is interpreted as 91.6971% of the variation in x is explained by y (and vice versa), and the remaining 8.3029% is noise caused by other uncaptured variables; 0.916971 is a pretty good coefficient of determination, showing that x and y explain each other's variance. But there could be other variables at play making up that remaining 0.083029. Remember, correlation does not equal causation, so there could be other variables contributing to the relationship we are seeing.

Correlation Is Not Causation!

It is important to note that while we put a lot of emphasis on measuring correlation and building metrics around it, please remember *correlation is not causation!* You probably have heard that mantra before, but I want to expand on why statisticians say it.

Just because we see a correlation between x and y does not mean x causes y. It could actually be y causes x! Or maybe there is a third uncaptured variable z that is causing x and y. It could be that x and y do not cause each other at all and the correlation is just coincidental, hence why it is important we measure the statistical significance.

Now I have a more pressing question for you. Can computers discern between correlation and causation? The answer is a resounding "NO!" Computers have concept of correlation but not causation. Let's say I load a dataset to scikit-learn showing gallons of water consumed and my water bill. My computer, or any program including scikit-learn, does not have any notion whether more water usage causes a higher bill or a higher bill causes more water usage. An AI system could easily conclude the latter, as nonsensical as that is. This is why many machine learning projects require a human in the loop to inject common sense.

In computer vision, this happens too. Computer vision often uses a regression on the numeric pixels to predict a category. If I train a computer vision system to recognize cows using pictures of cows, it can easily make correlations with the field rather than the cows. Therefore, if I show a picture of an empty field, it will label the grass as cows! This again is because computers have no concept of causality (the cow shape should cause the label "cow") and instead get lost in correlations we are not interested in.

Standard Error of the Estimate

One way to measure the overall error of a linear regression is the *SSE*, or *sum of squared error*. We learned about this earlier where we squared each residual and summed them. If $\widehat{y}$ (pronounced "y-hat") is each predicted value from the line and y represents each actual y-value from the data, here is the calculation:

$$SSE = \sum (y - \widehat{y})^2$$

However, all of these squared values are hard to interpret so we can use some square root logic to scale things back into their original units. We will also average all of them, and this is what the *standard error of the estimate (S_e)* does. If n is the number of data points, Example 5-19 shows how we calculate the standard error S_e in Python.

$$S_e = \sqrt{\frac{\Sigma(y - \hat{y})^2}{n - 2}}$$

Example 5-19. Calculating the standard error of the estimate

```
Here is how we calculate it in Python:

import pandas as pd
from math import sqrt

# Load the data
points = list(pd.read_csv('https://bit.ly/2KF29Bd', delimiter=",").itertuples())

n = len(points)

# Regression line
m = 1.939
b = 4.733

# Calculate Standard Error of Estimate
S_e = sqrt((sum((p.y - (m*p.x +b))**2 for p in points))/(n-2))

print(S_e)
# 1.87406793500129
```

Why $n - 2$ instead of $n - 1$, like we did in so many variance calculations in Chapter 3? Without going too deep into mathematical proofs, this is because a linear regression has two variables, not just one, so we have to increase the uncertainty by one more in our degrees of freedom.

You will notice the standard error of the estimate looks remarkably similar to the standard deviation we studied in Chapter 3. This is not by accident. That is because it is the standard deviation for a linear regression.

Prediction Intervals

As mentioned earlier, our data in a linear regression is a sample from a population. Therefore, our regression is only as good as our sample. Our linear regression line also has a normal distribution running along it. Effectively, this makes each predicted y-value a sample statistic just like the mean. As a matter of fact, the "mean" is shifting along the line.

Remember when we talked statistics in Chapter 2 about variance and standard deviation? The concepts apply here too. With a linear regression, we hope that data follows a normal distribution in a linear fashion. A regression line serves as the shifting "mean" of our bell curve, and the spread of the data around the line reflects the variance/standard deviation, as shown in Figure 5-16.

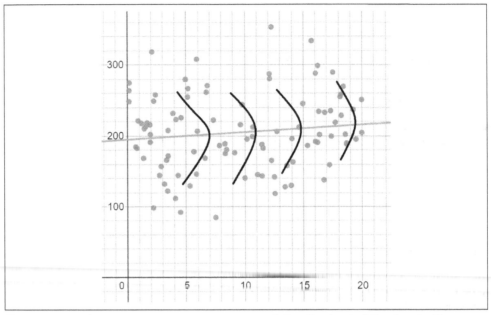

Figure 5-16. A linear regression assumes a normal distribution is following the line

When we have a normal distribution following a linear regression line, we have not just one variable but a second one steering a distribution as well. There is a confidence interval around each y prediction, and this is known as a *prediction interval*.

Let's bring back some context with our veterinary example, estimating the age of a dog and number of vet visits. I want to know the prediction interval for number of vet visits with 95% confidence for a dog that is 8.5 years old. What this prediction interval looks like is shown in Figure 5-17. We are 95% confident that an 8.5 year old dog will have between 16.462 and 25.966 veterinary visits.

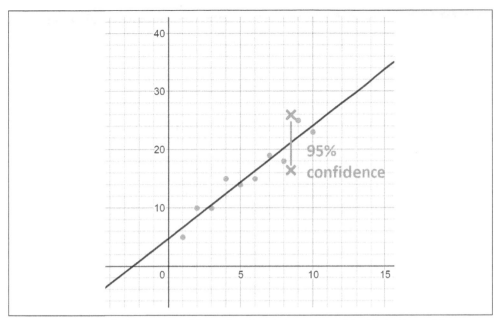

Figure 5-17. A prediction interval for a dog that is 8.5 years old with 95% confidence

How do we calculate this? We need to get the margin of error and plus/minus it around the predicted y-value. It's a beastly equation that involves a critical value from the T-distribution as well as the standard error of the estimate. Let's take a look:

$$E = t_{.025} * S_e * \sqrt{1 + \frac{1}{n} + \frac{n(x_0 - \bar{x})^2}{n(\Sigma x^2) - (\Sigma x)^2}}$$

The x-value we are interested in is specified as x_0, which in this case is 8.5. Here is how we solve this in Python, shown in Example 5-20.

Example 5-20. Calculating a prediction interval of vet visits for a dog that's 8.5 years old

```
import pandas as pd
from scipy.stats import t
from math import sqrt

# Load the data
points = list(pd.read_csv('https://bit.ly/2KF29Bd', delimiter=",").itertuples())

n = len(points)

# Linear Regression Line
m = 1.939
```

```
b = 4.733

# Calculate Prediction Interval for x = 8.5
x_0 = 8.5
x_mean = sum(p.x for p in points) / len(points)

t_value = t(n - 2).ppf(.975)

standard_error = sqrt(sum((p.y - (m * p.x + b)) ** 2 for p in points) / (n - 2))

margin_of_error = t_value * standard_error * \
                  sqrt(1 + (1 / n) + (n * (x_0 - x_mean) ** 2) / \
                      (n * sum(p.x ** 2 for p in points) - \
                          sum(p.x for p in points) ** 2))

predicted_y = m*x_0 + b

# Calculate prediction interval
print(predicted_y - margin_of_error, predicted_y + margin_of_error)
# 16.462516875955465 25.966483124044537
```

Oy vey! That's a lot of calculation, and unfortunately, SciPy and other mainstream data science libraries do not do this. But if you are inclined to statistical analysis, this is very useful information. We not only create a prediction based on a linear regression (e.g., a dog that's 8.5 years old will have 21.2145 vet visits), but we also are actually able to say something much less absolute: there's a 95% probability an 8.5 year old dog will visit the vet between 16.46 and 25.96 times. Brilliant, right? And it's a much safer claim because it captures a range rather than a single value, and thus accounts for uncertainty.

Confidence Intervals for the Parameters

When you think about it, the linear regression line itself is a sample statistic, and there is a linear regression line for the whole population we are trying to infer. This means parameters like m and b have their own distributions, and we can model confidence intervals around m and b individually to reflect the population's slope and y-intercept. This is beyond the scope of this book but it's worth noting this type of analysis can be done.

You can find resources to do these calculations from scratch but it might just be easier to use Excel's regression tools or libraries for Python.

Train/Test Splits

This analysis I just did with the correlation coefficient, statistical significance, and coefficient of determination unfortunately is not always done by practitioners. Sometimes they are dealing with so much data they do not have the time or technical ability to do so. For example a 128 × 128 pixel image is at least 16,384 variables. Do you have time to do statistical analysis on each of those pixel variables? Probably not! Unfortunately, this leads many data scientists to not learn these statistical metrics at all.

In an obscure online forum (*http://disq.us/p/1jas3zg*), I once read a post saying statistical regression is a scalpel, while machine learning is a chainsaw. When operating with massive amounts of data and variables, you cannot sift through all of that with a scalpel. You have to resort to a chainsaw and while you lose explainability and precision, you at least can scale to make broader predictions on more data. That being said, statistical concerns like sampling bias and overfitting do not go away. But there are a few practices that can be employed for quick validation.

Why Are There No Confidence Intervals and P-Values in scikit-learn?

Scikit-learn does not support confidence intervals and P-values, as these two techniques are open problems for higher-dimensional data. This only emphasizes the gap between statisticians and machine learning practitioners. As one of the maintainers of scikit-learn, Gael Varoquaux, said, "In general computing correct p-values requires assumptions on the data that are not met by the data used in machine learning (no multicollinearity, enough data compared to the dimensionality)....P-values are something that are expected to be well checked (they are a guard in medical research). Implementing them is asking for trouble....We can give p-values only in very narrow settings [with few variables]."

If you want to go down the rabbit hole, there are some interesting discussions on GitHub:

- *https://github.com/scikit-learn/scikit-learn/issues/6773*
- *https://github.com/scikit-learn/scikit-learn/issues/16802*

As mentioned before, statsmodel (*https://oreil.ly/8oEHo*) is a library that provides helpful tools for statistical analysis. Just know it will likely not scale to larger-dimensional models because of the aforementioned reasons.

A basic technique machine learning practitioners use to mitigate overfitting is a practice called the *train/test split*, where typically 1/3 of the data is set aside for testing and the other 2/3 is used for training (other ratios can be used as well). The *training dataset* is used to fit the linear regression, while the *testing dataset* is used to measure the linear regression's performance on data it has not seen before. This technique is generally used for all supervised machine learning, including logistic regression and neural networks. Figure 5-18 shows a visualization of how we break up our data into 2/3 for training and 1/3 for testing.

This Is a Small Dataset

As we will learn later, there are other ways to split a training/testing dataset than 2/3 and 1/3. If you have a dataset this small, you will probably be better off with 9/10 and 1/10 paired with cross-validation, or even just leave-one-out cross-validation. See "Do Train/Test Splits Have to be Thirds?" on page 189 to learn more.

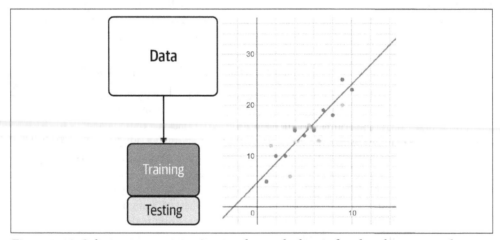

Figure 5-18. Splitting into training/testing data—the line is fitted to the training data (dark blue) using least squares, while the testing data (light red) is analyzed afterward to see how off the predictions are on data that was not seen before

Example 5-21 shows how to perform a train/test split using scikit-learn, where 1/3 of the data is set aside for testing and the other 2/3 is used for training.

Training Is Fitting a Regression

Remember that "fitting" a regression is synonymous with "training." The latter word is used by machine learning practitioners.

Example 5-21. Doing a train/test split on linear regression

```python
import pandas as pd
from sklearn.linear_model import LinearRegression
from sklearn.model_selection import train_test_split

# Load the data
df = pd.read_csv('https://bit.ly/3cIH97A', delimiter=",")

# Extract input variables (all rows, all columns but last column)
X = df.values[:, :-1]

# Extract output column (all rows, last column)
Y = df.values[:, -1]

# Separate training and testing data
# This leaves a third of the data out for testing
X_train, X_test, Y_train, Y_test = train_test_split(X, Y, test_size=1/3)

model = LinearRegression()
model.fit(X_train, Y_train)
result = model.score(X_test, Y_test)
print("r^2: %.3f" % result)
```

Notice that the `train_test_split()` will take our dataset (X and Y columns), shuffle it, and then return our training and testing datasets based on our testing-dataset size. We use the `LinearRegression`'s `fit()` function to fit on the training datasets `X_train` and `Y_train`. Then we use the `score()` function on the testing datasets `X_test` and `Y_test` to evaluate the r^2, giving us a sense how the regression performs on data it has not seen before. The higher the r^2 is for our testing dataset, the better. Having that higher number indicates the regression performs well on data it has not seen before.

Using R-Square for Testing

Note the r^2 is calculated a little bit differently here as we have a predefined linear regression from training. We are comparing the testing data to a regression line built from training data. The objective is still the same: being closer to 1.0 is desirable to show the regression correlation is strong even with the training data, while being closer to 0.0 indicates the testing dataset performs poorly. Here is how it is calculated, where y_i is each actual y-value, $\widehat{y_i}$ is each predicted y-value, and $\bar{y}$ is the average y-value for all data points:

$$r^2 = 1 - \frac{\Sigma(y_i - \widehat{y_i})^2}{\Sigma(y_i - \bar{y})^2}$$

Figure 5-19 shows different r^2 values from different linear regressions.

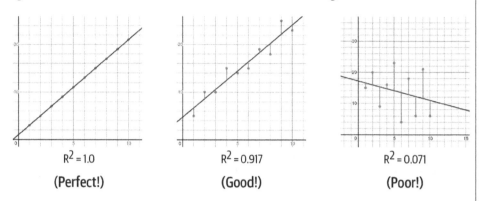

$R^2 = 1.0$

(Perfect!)

$R^2 = 0.917$

(Good!)

$R^2 = 0.071$

(Poor!)

Figure 5-19. An r^2 applied to different testing datasets with a trained linear regression

We can also alternate the testing dataset across each 1/3 fold. This is known as cross-validation and is often considered the gold standard of validation techniques. Figure 5-20 shows how each 1/3 of the data takes a turn being the testing dataset.

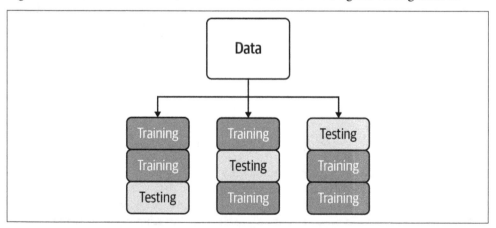

Figure 5-20. A visualization of cross-validation with three folds

The code in Example 5-22 shows a cross-validation performed across three folds, and then the scoring metric (in this case the mean sum of squares [MSE]) is averaged alongside its standard deviation to show how consistently each test performed.

Example 5-22. Using three-fold cross-validation for a linear regression

```
import pandas as pd
from sklearn.linear_model import LinearRegression
from sklearn.model_selection import KFold, cross_val_score

df = pd.read_csv('https://bit.ly/3cIH97A', delimiter=",")

# Extract input variables (all rows, all columns but last column)
X = df.values[:, :-1]

# Extract output column (all rows, last column)\
Y = df.values[:, -1]

# Perform a simple linear regression
kfold = KFold(n_splits=3, random_state=7, shuffle=True)
model = LinearRegression()
results = cross_val_score(model, X, Y, cv=kfold)
print(results)
print("MSE: mean=%.3f (stdev-%.3f)" % (results.mean(), results.std()))
```

Do Train/Test Splits Have to be Thirds?

We do not have to fold our data by thirds. You can use *k-fold validation* to split on any proportion. Typically, 1/3, 1/5, or 1/10 is used for the proportion of testing data, but 1/3 is the most common.

Generally, the *k* you choose has to result in a testing dataset that has a large enough sample for the problem. You also need enough alternating/shuffled testing datasets to provide a fair estimate of performance on data that was not seen before. Modestly sized datasets can use *k* values of 3, 5, or 10. *Leave-one-out cross-validation (LOOCV)* will alternate each individual data record as the only sample in the testing dataset, and this can be helpful when the whole dataset is small.

When you get concerned about variance in your model, one thing you can do, rather than a simple train/test split or cross-validation, is use *random-fold validation* to repeatedly shuffle and train/test split your data an unlimited number of times and aggregate the testing results. In Example 5-23 there are 10 iterations of randomly sampling 1/3 of the data for testing and the other 2/3 for training. Those 10 testing results are then averaged alongside their standard deviations to see how consistently the testing datasets perform.

What's the catch? It's computationally very expensive as we are training the regression many times.

Example 5-23. Using a random-fold validation for a linear regression

```
import pandas as pd
from sklearn.linear_model import LinearRegression
from sklearn.model_selection import cross_val_score, ShuffleSplit

df = pd.read_csv('https://bit.ly/38XwbeB', delimiter=",")

# Extract input variables (all rows, all columns but last column)
X = df.values[:, :-1]

# Extract output column (all rows, last column)\
Y = df.values[:, -1]

# Perform a simple linear regression
kfold = ShuffleSplit(n_splits=10, test_size=.33, random_state=7)
model = LinearRegression()
results = cross_val_score(model, X, Y, cv=kfold)

print(results)
print("mean=%.3f (stdev-%.3f)" % (results.mean(), results.std()))
```

So when you are crunched for time or your data is too voluminous to statistically analyze, a train/test split is going to provide a way to measure how well your linear regression will perform on data it has not seen before.

Train/Test Splits Are Not Guarantees

It is important to note that just because you apply machine learning best practices of splitting your training and testing data, it does not mean your model is going to perform well. You can easily overtune your model and p-hack your way into a good test result, only to find it does not work well out in the real world. This is why holding out another dataset called the *validation set* is sometimes necessary, especially if you are comparing different models or configurations. That way, your tweaks on the training data to get better performance on the testing data do not leak info into the training. You can use the validation dataset as one last stopgap to see if p-hacking caused you to overfit to your testing dataset.

Even then, your whole dataset (including training, testing, and validation) could have been biased to begin with, and no split is going to mitigate that. Andrew Ng discussed this as a large problem with machine learning during his Q&A with DeepLearning.AI and Stanford HAI (*https://oreil.ly/x23SJ*). He walked through an example showing why machine learning has not replaced radiologists.

Multiple Linear Regression

We put almost exclusive focus on doing linear regression on one input variable and one output variable in this chapter. However, the concepts we learned here should largely apply to multivariable linear regression. Metrics like r^2, standard error, and confidence intervals can be used but it gets harder with more variables. Example 5-24 is an example of a linear regression with two input variables and one output variable using scikit-learn.

Example 5-24. A linear regression with two input variables

```
import pandas as pd
from sklearn.linear_model import LinearRegression

# Load the data
df = pd.read_csv('https://bit.ly/2X1HWH7', delimiter=",")

# Extract input variables (all rows, all columns but last column)
X = df.values[:, :-1]

# Extract output column (all rows, last column)\
Y = df.values[:, -1]

# Training
fit = LinearRegression().fit(X, Y)

# Print coefficients
print("Coefficients = {0}".format(fit.coef_))
print("Intercept = {0}".format(fit.intercept_))
print("z = {0} + {1}x + {2}y".format(fit.intercept_, fit.coef_[0], fit.coef_[1]))
```

There is a degree of precariousness when a model becomes so inundated with variables it starts to lose explainability, and this is when machine learning practices start to come in and treat the model as as black box. I hope that you are convinced statistical concerns do not go away, and data becomes increasingly sparse the more variables you add. But if you step back and analyze the relationships between each pair of variables using a correlation matrix, and seek understanding on how each pair of variables interact, it will help your efforts to create a productive machine learning model.

Conclusion

We covered a lot in this chapter. We attempted to go beyond a cursory understanding of linear regression and making train/test splits our only validation. I wanted to show you both the scalpel (statistics) and the chainsaw (machine learning) so you can judge which is best for a given problem you encounter. There are a lot of metrics and analysis methods available in linear regression alone, and we covered a number of them to understand whether a linear regression is reliable for predictions. You may find yourself in a position to do regressions as broad approximations or meticulously analyze and comb your data using statistical tools. Which approach you use is situational, and if you want to learn more about statistical tools available for Python, check out the statsmodel library (*https://oreil.ly/8oEHo*).

In Chapter 6 covering logistic regression, we will revisit the r^2 and statistical significance. I hope that this chapter convinced you there are ways to analyze data meaningfully, and the investment can make the difference in a successful project.

Exercises

A dataset of two variables, x and y, is provided here (*https://bit.ly/3C8JzrM*).

1. Perform a simple linear regression to find the m and b values that minimizes the loss (sum of squares).

2. Calculate the correlation coefficient and statistical significance of this data (at 95% confidence). Is the correlation useful?

3. If I predict where $x = 50$, what is the 95% prediction interval for the predicted value of y?

4. Start your regression over and do a train/test split. Feel free to experiment with cross-validation and random-fold validation. Does the linear regression perform well and consistently on the testing data? Why or why not?

Answers are in Appendix B.

## CHAPTER 6

Logistic Regression and Classification

In this chapter we are going to cover *logistic regression*, a type of regression that predicts a probability of an outcome given one or more independent variables. This in turn can be used for *classification*, which is predicting categories rather than real numbers as we did with linear regression.

We are not always interested in representing variables as *continuous*, where they can represent an infinite number of real decimal values. There are situations where we would rather variables be *discrete*, or representative of whole numbers, integers, or booleans (1/0, true/false). Logistic regression is trained on an output variable that is discrete (a binary 1 or 0) or a categorical number (which is a whole number). It does output a continuous variable in the form of probability, but that can be converted into a discrete value with a threshold.

Logistic regression is easy to implement and fairly resilient against outliers and other data challenges. Many machine learning problems can best be solved with logistic regression, offering more practicality and performance than other types of supervised machine learning.

Just like we did in Chapter 5 when we covered linear regression, we will attempt to walk the line between statistics and machine learning, using tools and analysis from both disciplines. Logistic regression will integrate many concepts we have learned from this book, from probability to linear regression.

Understanding Logistic Regression

Imagine there was a small industrial accident and you are trying to understand the impact of chemical exposure. You have 11 patients who were exposed for differing numbers of hours to this chemical (please note this is fabricated data). Some have shown symptoms (value of 1) and others have not shown symptoms (value of 0).

Let's plot them in Figure 6-1, where the x-axis is hours of exposure and the y-axis is whether or not (1 or 0) they have showed symptoms.

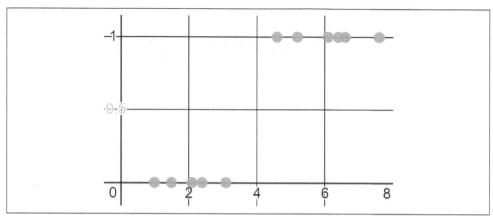

Figure 6-1. Plotting whether patients showed symptoms (1) or not (0) over x hours of exposure

At what length of time do patients start showing symptoms? Well it is easy to see at almost four hours, we immediately transition from patients not showing symptoms (0) to showing symptoms (1). In Figure 6-2, we see the same data with a predictive curve.

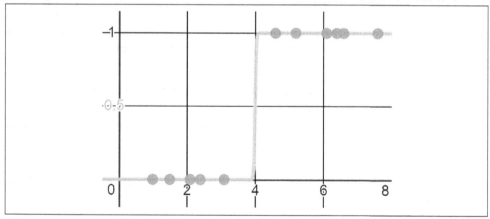

Figure 6-2. After four hours, we see a clear jump where patients start showing symptoms

Doing a cursory analysis on this sample, we can say that there is nearly 0% probability a patient exposed for fewer than four hours will show symptoms, but there is 100% probability for greater than four hours. Between these two groups, there is an immediate jump to showing symptoms at approximately four hours.

Of course, nothing is ever this clear-cut in the real world. Let's say you gathered more data, where the middle of the range has a mix of patients showing symptoms versus not showing symptoms as shown in Figure 6-3.

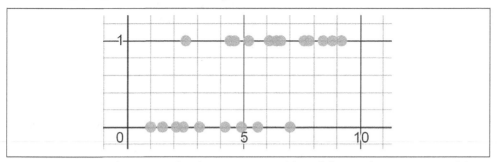

Figure 6-3. A mix of patients who show symptoms (1) and do not show symptoms (0) exists in the middle

The way to interpret this is the probability of patients showing symptoms gradually increases with each hour of exposure. Let's visualize this with a *logistic function*, or an S-shaped curve where the output variable is squeezed between 0 and 1, as shown in Figure 6-4.

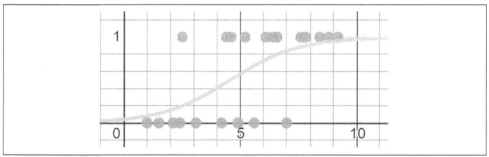

Figure 6-4. Fitting a logistic function to the data

Because of this overlap of points in the middle, there is no distinct cutoff when patients show symptoms but rather a gradual transition from 0% probability to 100% probability (0 and 1). This example demonstrates how a *logistic regression* results in a curve indicating a probability of belonging to the true category (a patient showed symptoms) across an independent variable (hours of exposure).

We can repurpose a logistic regression to not just predict a probability for given input variables but also add a threshold to predict whether it belongs to that category. For example, if I get a new patient and find they have been exposed for six hours, I predict a 71.1% chance they will show symptoms as traced in Figure 6-5. If my threshold is at least 50% probability to show symptoms, I will simply classify that the patient will show symptoms.

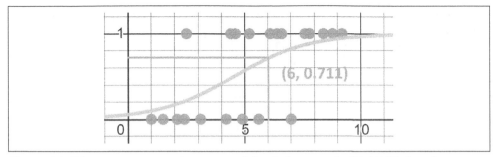

Figure 6-5. We can expect a patient exposed for six hours to be 71.1% likely to have symptoms, and because that's greater than a threshold of 50% we predict they will show symptoms

Performing a Logistic Regression

So how do we perform a logistic regression? Let's first take a look at the logistic function and explore the math behind it.

Logistic Function

The *logistic function* is an S-shaped curve (also known as a *sigmoid curve*) that, for a given set of input variables, produces an output variable between 0 and 1. Because the output variable is between 0 and 1 it can be used to represent a probability.

Here is the logistic function that outputs a probability y for one input variable x:

$$y = \frac{1.0}{1.0 + e^{-(\beta_0 + \beta_1 x)}}$$

Note this formula uses Euler's number e, which we covered in Chapter 1. The x variable is the independent/input variable. β_0 and β_1 are the coefficients we need to solve for.

β_0 and β_1 are packaged inside an exponent resembling a linear function, which you may recall looks identical to $y = mx + b$ or $y = \beta_0 + \beta_1 x$. This is not a coincidence; logistic regression actually has a close relationship to linear regression, which we will discuss later in this chapter. β_0 indeed is the intercept (which we call b in a simple linear regression) and β_1 is the slope for x (which we call m in a simple linear regression). This linear function in the exponent is known as the log-odds function, but for now just know this whole logistic function produces this S-shaped curve we need to output a shifting probability across an x-value.

To declare the logistic function in Python, use the `exp()` function from the `math` package to declare the e exponent as shown in Example 6-1.

Example 6-1. The logistic function in Python for one independent variable

```
import math

def predict_probability(x, b0, b1):
    p = 1.0 / (1.0 + math.exp(-(b0 + b1 * x)))
    return p
```

Let's plot to see what it looks like, and assume $B_0 = -2.823$ and $B_1 = 0.62$. We will use SymPy in Example 6-2 and the output graph is shown in Figure 6-6.

Example 6-2. Using SymPy to plot a logistic function

```
from sympy import *
b0, b1, x = symbols('b0 b1 x')

p = 1.0 / (1.0 + exp(-(b0 + b1 * x)))

p = p.subs(b0,-2.823)
p = p.subs(b1, 0.620)
print(p)

plot(p)
```

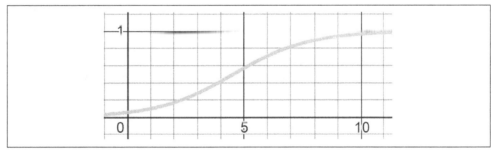

Figure 6-6. A logistic function

In some textbooks, you may alternatively see the logistic function declared like this:

$$p = \frac{e^{\beta_0 + \beta_1 x}}{1 + e^{\beta_0 + \beta_1 x}}$$

Do not fret about it, because it is the same function, just algebraically expressed differently. Note like linear regression we can also extend logistic regression to more than one input variable $(x_1, x_2, \ldots x_n)$, as shown in this formula. We just add more β_x coefficients:

$$p = \frac{1}{1 + e^{-(\beta_0 + \beta_1 x_1 + \beta_2 x_2 + \ldots \beta_n x_n)}}$$

Fitting the Logistic Curve

How do you fit the logistic curve to a given training dataset? First, the data can have any mix of decimal, integer, and binary variables, but the output variable must be binary (0 or 1). When we actually do prediction, the output variable will be between 0 and 1, resembling a probability.

The data provides our input and output variable values, but we need to solve for the β_0 and β_1 coefficients to fit our logistic function. Recall how we used least squares in Chapter 5. However, this does not apply here. Instead we use *maximum likelihood estimation*, which, as the name suggests, maximizes the likelihood a given logistic curve would output the observed data.

To calculate the maximum likelihood estimation, there really is no closed form equation like in linear regression. We can still use gradient descent, or have a library do it for us. Let's cover both of these approaches starting with the library SciPy.

Using scikit-learn

The nice thing about SciPy is the models often have a standardized set of functions and APIs, meaning in many cases you can copy/paste your code and can then reuse it between models. In Example 6-3 you will see a logistic regression performed on our patient data. If you compare it to our linear regression code in Chapter 5, you will see it has nearly identical code in importing, separating, and fitting our data. The main difference is I use a `LogisticRegression()` for my model instead of a `LinearRegression()`.

Example 6-3. Using a plain logistic regression in SciPy

```
import pandas as pd
from sklearn.linear_model import LogisticRegression

# Load the data
df = pd.read_csv('https://bit.ly/33ebs2R', delimiter=",")

# Extract input variables (all rows, all columns but last column)
X = df.values[:, :-1]

# Extract output column (all rows, last column)
Y = df.values[:, -1]

# Perform logistic regression
# Turn off penalty
model = LogisticRegression(penalty=None)
```

```
model.fit(X, Y)

# print beta1
print(model.coef_.flatten()) # 0.69267212

# print beta0
print(model.intercept_.flatten()) # -3.17576395
```

Making Predictions

To make specific predictions, use the predict() and predict_prob() functions on the model object in scikit-learn, whether it is a LogisticRegression or any other type of classification model. The predict() function will predict a specific class (e.g., True 1.0 or False 1.0) while the predict_prob() will output probabilities for each class.

After running the model in scikit-learn, I get a logistic regression where $\beta_0 = -3.17576395$ and $\beta_1 = 0.69267212$. When I plot this, it should look pretty good as shown in Figure 6-7.

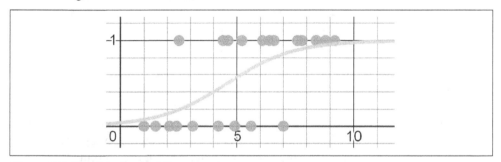

Figure 6-7. Plotting the logistic regression

There are a couple of things to note here. When I created the LogisticRegression() model, I specified no penalty argument, which chooses a regularization technique like l1 or l2. While this is beyond the scope of this book, I have included brief insights in the following note "Learning About scikit-learn Parameters" so that you have helpful references on hand.

Finally, I am going to flatten() the coefficient and intercept, which come out as multidimensional matrices but with one element. *Flattening* means collapsing a matrix of numbers into lesser dimensions, particularly when there are fewer elements than there are dimensions. For example, I use flatten() here to take a single number nested into a two-dimensional matrix and pull it out as a single value. I then have my β_0 and β_1 coefficients.

Learning About scikit-learn Parameters

scikit-learn offers a lot of options in its regression and classification models. Unfortunately, there is not enough bandwidth or pages to cover them as this is not a book focusing exclusively on machine learning.

However, the scikit-learn docs are well-written and the page on logistic regression is found here (*https://oreil.ly/eL8hZ*).

If a lot of terms are unfamiliar, such as regularization and l1 and l2 penalties, there are other great O'Reilly books exploring these topics. One of the more helpful texts I have found is *Hands-On Machine Learning with Scikit-Learn, Keras, and TensorFlow* by Aurélien Géron.

Using Maximum Likelihood and Gradient Descent

As I have done throughout this book, I aim to provide insights on building techniques from scratch even if libraries can do it for us. There are several ways to fit a logistic regression ourselves, but all methods typically turn to maximum likelihood estimation (MLE). MLE maximizes the likelihood a given logistic curve would output the observed data. It is different than sum of squares, but we can still apply gradient descent or stochastic gradient descent to solve it.

I'll try to streamline the mathematical jargon and minimize the linear algebra here. Essentially, the idea is to find the β_0 and β_1 coefficients that bring our logistic curve to those points as closely as possible, indicating it is most likely to have produced those points. If you recall from Chapter 2 when we studied probability, we combine probabilities (or likelihoods) of multiple events by multiplying them together. In this application, we are calculating the likelihood we would see all these points for a given logistic regression curve.

Applying the idea of joint probabilities, each patient has a likelihood they would show symptoms *based on the fitted logistic function* as shown in Figure 6-8.

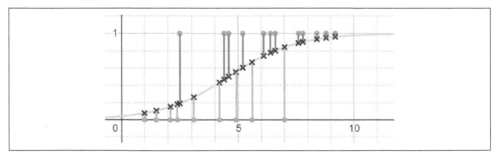

Figure 6-8. Every input value has a corresponding likelihood on the logistic curve

We fetch each likelihood off the logistic regression curve above or below each point. If the point is below the logistic regression curve, we need to subtract the resulting probability from 1.0 because we want to maximize the false cases too.

Given coefficients $\beta_0 = -3.17576395$ and $\beta_1 = 0.69267212$, Example 6-4 shows how we calculate the joint likelihood for this data in Python.

Example 6-4. Calculating the joint likelihood of observing all the points for a given logistic regression

```
import math
import pandas as pd

patient_data = pd.read_csv('https://bit.ly/33ebs2R', delimiter=",").itertuples()

b0 = -3.17576395
b1 = 0.69267212

def logistic_function(x):
    p = 1.0 / (1.0 + math.exp(-(b0 + b1 * x)))
    return p

# Calculate the joint likelihood
joint_likelihood = 1.0

for p in patient_data:
    if p.y == 1.0:
        joint_likelihood *= logistic_function(p.x)
    elif p.y == 0.0:
        joint_likelihood *= (1.0 - logistic_function(p.x))

print(joint_likelihood) # 4.7911180221699105e-05
```

Here's a mathematical trick we can do to compress that if expression. As we covered in Chapter 1, when you set any number to the power of 0 it will always be 1. Take a look at this formula and note the handling of true (1) and false (0) cases in the exponents:

$$\text{joint likelihood} = \prod_{i=1}^{n} \left(\frac{1.0}{1.0 + e^{-(\beta_0 + \beta_1 x_i)}} \right)^{y_i} \times \left(1.0 - \frac{1.0}{1.0 + e^{-(\beta_0 + \beta_1 x_i)}} \right)^{1.0 - y_i}$$

To do this in Python, compress everything inside that for loop into Example 6-5.

Example 6-5. Compressing the joint likelihood calculation without an `if` expression

```
for p in patient_data:
    joint_likelihood *= logistic_function(p.x) ** p.y * \
                        (1.0 - logistic_function(p.x)) ** (1.0 - p.y)
```

What exactly did I do? Notice that there are two halves to this expression, one for when $y = 1$ and the other where $y = 0$. When any number is raised to exponent 0, it will result in 1. Therefore, whether y is 1 or 0, it will cause the opposite condition on the other side to evaluate to 1 and have no effect in multiplication. We get to express our `if` expression but do it completely in a mathematical expression. We cannot do derivatives on expressions that use `if`, so this will be helpful.

Note that computers can get overwhelmed multiplying several small decimals together, known as *floating point underflow*. This means that as decimals get smaller and smaller, which can happen in multiplication, the computer runs into limitations keeping track of that many decimal places. There is a clever mathematical hack to get around this. You can take the `log()` of each decimal you are multiplying and instead add them together. This is thanks to the additive properties of logarithms we covered in Chapter 1. This is more numerically stable, and you can then call the `exp()` function to convert the total sum back to get the product.

Let's revise our code to use logarithmic addition instead of multiplication (see Example 6-6). Note that the `log()` function will default to base e and while any base technically works, this is preferable because e^x is the derivative of itself and will computationally be more efficient.

Example 6-6. Using logarithmic addition

```
# Calculate the joint likelihood
joint_likelihood = 0.0

for p in patient_data:
    joint_likelihood += math.log(logistic_function(p.x) ** p.y * \
                        (1.0 - logistic_function(p.x)) ** (1.0 - p.y))

joint_likelihood = math.exp(joint_likelihood)
```

To express the preceding Python code in mathematical notation:

$$\text{joint likelihood} = \sum_{i=1}^{n} log\left(\left(\frac{1.0}{1.0 + e^{-(\beta_0 + \beta_1 x_i)}}\right)^{y_i} \times \left(1.0 - \frac{1.0}{1.0 + e^{-(\beta_0 + \beta_1 x_i)}}\right)^{1.0 - y_i}\right)$$

Would you like to calculate the partial derivatives for β_0 and β_1 in the preceding expression? I didn't think so. It's a beast. Goodness, expressing that function in SymPy alone is a mouthful! Look at this in Example 6-7.

Example 6-7. Expressing a joint likelihood for logistic regression in SymPy

```
joint_likelihood = Sum(log((1.0 / (1.0 + exp(-(b + m * x(i)))))**y(i) * \
        (1.0 - (1.0 / (1.0 + exp(-(b + m * x(i))))))**(1-y(i))), (i, 0, n))
```

So let's just allow SymPy to do the partial derivatives for us, for β_0 and β_1 respectively. We will then immediately compile and use them for gradient descent, as shown in Example 6-8.

Example 6-8. Using gradient descent on logistic regression

```
from sympy import *
import pandas as pd

points = list(pd.read_csv("https://tinyurl.com/y2cocoo7").itertuples())

b1, b0, i, n = symbols('b1 b0 i n')
x, y = symbols('x y', cls=Function)
joint_likelihood = Sum(log((1.0 / (1.0 + exp(-(b0 + b1 * x(i))))) ** y(i) \
        * (1.0 - (1.0 / (1.0 + exp(-(b0 + b1 * x(i)))))) ** (1 - y(i))), (i, 0, n))

# Partial derivative for m, with points substituted
d_b1 = diff(joint_likelihood, b1) \
                .subs(n, len(points) - 1).doit() \
                .replace(x, lambda i: points[i].x) \
                .replace(y, lambda i: points[i].y)

# Partial derivative for m, with points substituted
d_b0 = diff(joint_likelihood, b0) \
                .subs(n, len(points) - 1).doit() \
                .replace(x, lambda i: points[i].x) \
                .replace(y, lambda i: points[i].y)

# compile using lambdify for faster computation
d_b1 = lambdify([b1, b0], d_b1)
d_b0 = lambdify([b1, b0], d_b0)

# Perform Gradient Descent
b1 = 0.01
b0 = 0.01
L = .01

for j in range(10_000):
    b1 += d_b1(b1, b0) * L
    b0 += d_b0(b1, b0) * L
```

```
print(b1, b0)
# 0.6926693075370812 -3.175751550409821
```

After calculating the partial derivatives for β_0 and β_1, we substitute the x- and y-values as well as the number of data points n. Then we use `lambdify()` to compile the derivative function for efficiency (it uses NumPy behind the scenes). After that, we perform gradient descent like we did in Chapter 5, but since we are trying to maximize rather than minimize, we add each adjustment to β_0 and β_1 rather than subtract like in least squares.

As you can see in Example 6-8, we got $\beta_0 = -3.17575$ and $\beta_1 = 0.692667$. This is highly comparable to the coefficient values we got in SciPy earlier.

As we learned to do in Chapter 5, we can also use stochastic gradient descent and only sample one or a handful of records on each iteration. This would extend the benefits of increasing computational speed and performance as well as prevent overfitting. It would be redundant to cover it again here, so we will keep moving on.

Multivariable Logistic Regression

Let's try an example that uses logistic regression on multiple input variables. Table 6-1 shows a sample of a few records from a fictitious dataset containing some employment-retention data (full dataset is here (*https://bit.ly/3aqsOMO*)).

Table 6-1. Sample of employment-retention data

SEX	AGE	PROMOTIONS	YEARS_EMPLOYED	DID_QUIT
1	32	3	7	0
1	34	2	5	0
1	29	2	5	1
0	42	4	10	0
1	43	4	10	0

There are 54 records in this dataset. Let's say we want to use it to predict whether other employees are going to quit and logistic regression can be utilized here (although none of this is a good idea, and I will elaborate why later). Recall we can support more than one input variable as shown in this formula:

$$y = \frac{1}{1 + e^{-(\beta_0 + \beta_1 x_1 + \beta_2 x_2 + \ldots \beta_n x_n)}}$$

I will create β coefficients for each of the variables `sex`, `age`, `promotions`, and `years_employed`. The output variable `did_quit` is binary, and that is going to drive the logistic regression outcome we are predicting. Because we are dealing with multiple dimensions, it is going to be hard to visualize the curvy hyperplane that is our logistic curve. So we will steer clear from visualization.

Let's make it interesting. We will use scikit-learn but make an interactive shell we can test employees with. Example 6-9 shows the code, and when we run it, a logistic regression will be performed, and then we can type in new employees to predict whether they quit or not. What can go wrong? Nothing, I'm sure. We are only making predictions on people's personal attributes and making decisions accordingly. I'm sure it will be fine.

(If it was not clear, I'm being very tongue in cheek).

Example 6-9. Doing a multivariable logistic regression on employee data

```python
import pandas as pd
from sklearn.linear_model import LogisticRegression

employee_data = pd.read_csv("https://tinyurl.com/y6r7qjrp")

# grab independent variable columns
inputs = employee_data.iloc[:, :-1]

# grab dependent "did_quit" variable column
output = employee_data.iloc[:, -1]

# build logistic regression
fit = LogisticRegression(penalty='none').fit(inputs, output)

# Print coefficients:
print("COEFFICIENTS: {0}".format(fit.coef_.flatten()))
print("INTERCEPT: {0}".format(fit.intercept_.flatten()))

# Interact and test with new employee data
def predict_employee_will_stay(sex, age, promotions, years_employed):
    prediction = fit.predict([[sex, age, promotions, years_employed]])
    probabilities = fit.predict_proba([[sex, age, promotions, years_employed]])
    if prediction == [[1]]:
        return "WILL LEAVE: {0}".format(probabilities)
    else:
        return "WILL STAY: {0}".format(probabilities)

# Test a prediction
while True:
    n = input("Predict employee will stay or leave {sex},
        {age},{promotions},{years employed}: ")
    (sex, age, promotions, years_employed) = n.split(",")
    print(predict_employee_will_stay(int(sex), int(age), int(promotions),
        int(years_employed)))
```

Figure 6-9 shows the result whether an employee is predicted to quit. The employee is a sex "1," age is 34, had 1 promotion, and has been at the company for 5 years. Sure enough, the prediction is "WILL LEAVE."

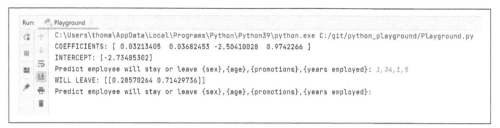

Figure 6-9. Making a prediction whether a 34-year-old employee with 1 promotion and 5 years, employment will quit

Note that the `predict_proba()` function will output two values, the first being the probability of 0 (false) and the second being 1 (true).

You will notice that the coefficients for `sex`, `age`, `promotions`, and `years_employed` are displayed in that order. By the weight of the coefficients, you can see that `sex` and `age` play very little role in the prediction (they both have a weight near 0). However, `promotions` and `years_employed` have significant weights of `-2.504` and `0.97`. Here's a secret with this toy dataset: I fabricated it so that an employee quits if they do not get a promotion roughly every two years. Sure enough, my logistic regression picked up this pattern and you can try it out with other employees as well. However, if you venture outside the ranges of data it was trained on, the predictions will likely start falling apart (e.g., if put in a 70-year-old employee who hasn't been promoted in three years, it's hard to say what this model will do since it has no data around that age).

Of course, real life is not always this clean. An employee who has been at a company for eight years and has never gotten a promotion is likely comfortable with their role and not leaving anytime soon. If that is the case, variables like age then might play a role and get weighted. Then of course we can get concerned about other relevant variables that are not being captured. See the following warning to learn more.

Be Careful Making Classifications on People!

A quick and surefire way to shoot yourself in the foot is to collect data on people and use it to make predictions haphazardly. Not only can data privacy concerns come about, but legal and PR issues can emerge if the model is found to be discriminatory. Input variables like race and gender can become weighted from machine learning training. After that, undesirable outcomes are inflicted on those demographics like not being hired or being denied loans. More extreme applications include being falsely flagged by surveillance systems or being denied criminal parole. Note too that seemingly benign variables like commute time have been found to correlate with discriminatory variables.

At the time of writing, a number of articles have been citing machine learning discrimination as an issue:

- Katyanna Quach, "Teen turned away from roller rink after AI wrongly identifies her as banned troublemaker" (*https://oreil.ly/boUcW*), *The Register*, July 16, 2021.
- Kashmir Hill, "Wrongfully Accused by an Algorithm" (*https://oreil.ly/dOJyI*), *New York Times*, June 24, 2020.

As data privacy laws continue to evolve, it is advisable to err on the side of caution and engineer personal data carefully. Think about what automated decisions will be propagated and how that can cause harm. Sometimes it is better to just leave a "problem" alone and keep doing it manually.

Finally, on this employee-retention example, think about where this data came from. Yes, I made up this dataset but in the real world you always want to question what process created the data. Over what period of time did this sample come from? How far back do we go looking for employees who quit? What constitutes an employee who stayed? Are they current employees at this point in time? How do we know they are not about to quit, making them a false negative? Data scientists easily fall into traps analyzing only what data says, but not questioning where it came from and what assumptions are built into it.

The best way to get answers to these questions is to understand what the predictions are being used for. Is it to decide when to give people promotions to retain them? Can this create a circular bias promoting people with a set of attributes? Will that bias be reaffirmed when those promotions start becoming the new training data?

These are all important questions, and perhaps even inconvenient ones that cause unwanted scope to creep into the project. If this scrutiny is not welcomed by your team or leadership on a project, consider empowering yourself with a different role where curiosity becomes a strength.

Understanding the Log-Odds

At this point, it is time to discuss the logistic regression and what it is mathematically made of. This can be a bit dizzying so take your time here. If you get overwhelmed, you can always revisit this section later.

Starting in the 1900s, it has always been of interest to mathematicians to take a linear function and scale its output to fall between 0 and 1, and therefore be useful for predicting probability. The log-odds, also called the logit function, lends itself to logistic regression for this purpose.

Remember earlier I pointed out the exponent value $\beta_0 + \beta_1 x$ is a linear function? Look at our logistic function again:

$$p = \frac{1.0}{1.0 + e^{-(\beta_0 + \beta_1 x)}}$$

This linear function being raised to e is known as the *log-odds* function, which takes the logarithm of the odds for the event of interest. Your response might be, "Wait, I don't see any log() or odds. I just see a linear function!" Bear with me, I will show the hidden math.

As an example, let's use our logistic regression from earlier where $B_0 = 3.17576395$ and $B_1 = 0.69267212$. What is the probability of showing symptoms after six hours, where $x = 6$? We already know how to do this: plug these values into our logistic function:

$$p = \frac{1.0}{1.0 + e^{-(-3.17576395 + 0.69267212(6))}} = 0.727161542928554$$

We plug in these values and output a probability of 0.72716. But let's look at this from an odds perspective. Recall in Chapter 2 we learned how to calculate odds from a probability:

$$\text{odds} = \frac{p}{1-p}$$

$$\text{odds} = \frac{.72716}{1 - .72716} = 2.66517246407876$$

So at six hours, a patient is 2.66517 times more likely to show symptoms than not show symptoms.

When we wrap the odds function in a natural logarithm (a logarithm with base e), we call this the *logit function*. The output of this formula is what we call the *log-odds*, named...shockingly...because we take the logarithm of the odds:

$$logit = log\left(\frac{p}{1-p}\right)$$

$$logit = log\left(\frac{.72716}{1-.72716}\right) = 0.98026877$$

Our log-odds at six hours is 0.9802687. What does this mean and why do we care? When we are in "log-odds land" it is easier to compare one set of odds against another. We treat anything greater than 0 as favoring odds an event will happen, whereas anything less than 0 is against an event. A log-odds of –1.05 is linearly the same distance from 0 as 1.05. In plain odds, though, the equivalents are 0.3499 and 2.857, respectively, which is not as interpretable. That is the convenience of log-odds.

Odds and Logs

Logarithms and odds have an interesting relationship. Odds are against an event when it is between 0.0 and 1.0, but anything greater than 1.0 favors the event and extends into positive infinity. This lack of symmetry is awkward. However, logarithms rescale an odds so that it is completely linear, where a log-odds of 0.0 means fair odds. A log-odds of –1.05 is linearly the same distance from 0 as 1.05, thus make comparing odds much easier.

Josh Starmer has a great video (*https://oreil.ly/VOH8w*) talking about this relationship between odds and logs.

Recall I said the linear function in our logistic regression formula $\beta_0 + \beta_1 x$ is our log-odds function. Check this out:

log-odds $= \beta_0 + \beta_1 x$
log-odds $= -3.17576395 + 0.69267212(6)$
log-odds $= 0.98026877$

It's the same value 0.98026877 as our previous calculation, the odds of our logistic regression at $x = 6$ and then taking the $log()$ of it! So what is the link? What ties all this together? Given a probability from a logistic regression p and input variable x, it is this:

$$log\left(\frac{p}{1-p}\right) = \beta_0 + \beta_1 x$$

Let's plot the log-odds line alongside the logistic regression, as shown in Figure 6-10.

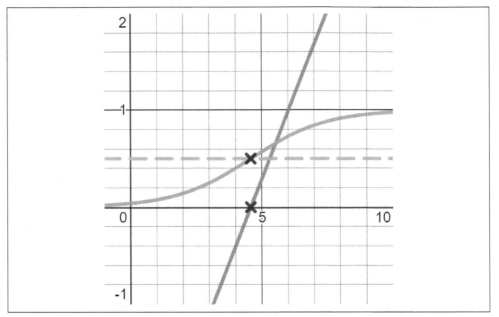

Figure 6-10. The log-odds line is converted into a logistic function that outputs a probability

Every logistic regression is actually backed by a linear function, and that linear function is a log-odds function. Note in Figure 6-10 that when the log-odds is 0.0 on the line, then the probability of the logistic curve is at 0.5. This makes sense because when our odds are fair at 1.0, the probability is going to be 0.50 as shown in the logistic regression, and the log-odds are going to be 0 as shown by the line.

Another benefit we get looking at the logistic regression from an odds perspective is we can compare the effect between one x-value and another. Let's say I want to understand how much my odds change between six hours and eight hours of exposure to the chemical. I can take the odds at six hours and then eight hours, and then ratio the two odds against each other in an *odds ratio*. This is not to be confused with a plain odds which, yes, is a ratio, but it is not an odds ratio.

Let's first find the probabilities of symptoms for six hours and eight hours, respectively:

$$p = \frac{1.0}{1.0 + e^{-(\beta_0 + \beta_1 x)}}$$

$$p_6 = \frac{1.0}{1.0 + e^{-(-3.17576395 + 0.69267212(6))}} = 0.727161542928554$$

$$p_8 = \frac{1.0}{1.0 + e^{-(-3.17576395 + 0.69267212(8))}} = 0.914167258137741$$

Now let's convert those into odds, which we will declare as o_x:

$$o = \frac{p}{1 - p}$$

$$o_6 = \frac{0.727161542928554}{1 - 0.727161542928554} = 2.66517246407876$$

$$o_8 = \frac{0.914167258137741}{1 - 0.914167258137741} = 10.6505657200694$$

Finally, set the two odds against each other as an odds ratio, where the odds for eight hours is the numerator and the odds for six hours is in the denominator. We get a value of approximately 3.996, meaning that our odds of showing symptoms increases by nearly a factor of four with an extra two hours of exposure:

$$\text{odds ratio} = \frac{10.6505657200694}{2.66517246407876} = 3.99620132040906$$

You will find this odds ratio value of 3.996 holds across any two-hour range, like 2 hours to 4 hours, 4 hours to 6 hours, 8 hours to 10 hours, and so forth. As long as it's a two-hour gap, you will find that odds ratio stays consistent. It will differ for other range lengths.

R-Squared

We covered quite a few statistical metrics for linear regression in Chapter 5, and we will try to do the same for logistic regression. We still worry about many of the same problems as in linear regression, including overfitting and variance. As a matter of fact, we can borrow and adapt several metrics from linear regression and apply them to logistic regression. Let's start with R^2.

Just like linear regression, there is an R^2 for a given logistic regression. If you recall from Chapter 5, the R^2 indicates how well a given independent variable explains a dependent variable. Applying this to our chemical exposure problem, it makes sense we want to measure how much chemical exposure hours explains showing symptoms.

There is not really a consensus on the best way to calculate the R^2 on a logistic regression, but a popular technique known as McFadden's Pseudo R^2 closely mimics the R^2 used in linear regression. We will use this technique in the following examples and here is the formula:

$$R^2 = \frac{(\text{log likelihood}) - (\text{log likelihood fit})}{(\text{log likelihood})}$$

We will learn how to calculate the "log likelihood fit" and "log likelihood" so we can calculate the R^2.

We cannot use residuals here like in linear regression, but we can project the outcomes back onto the logistic curve as shown in Figure 6-11, and look up their corresponding likelihoods between 0.0 and 1.0.

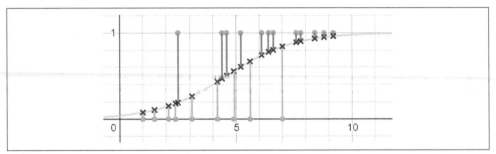

Figure 6-11. Projecting the output values back onto the logistic curve

We can then take the `log()` of each of those likelihoods and sum them together. This will be the log likelihood of the fit (Example 6-10). Just like we did calculating maximum likelihood, we will convert the "false" likelihoods by subtracting from 1.0.

Example 6-10. Calculating the log likelihood of the fit

```
from math import log, exp
import pandas as pd

patient_data = pd.read_csv('https://bit.ly/33ebs2R', delimiter=",").itertuples()

b0 = -3.17576395
b1 = 0.69267212

def logistic_function(x):
    p = 1.0 / (1.0 + exp(-(b0 + b1 * x)))
    return p
```

```
# Sum the log-likelihoods
log_likelihood_fit = 0.0

for p in patient_data:
    if p.y == 1.0:
        log_likelihood_fit += log(logistic_function(p.x))
    elif p.y == 0.0:
        log_likelihood_fit += log(1.0 - logistic_function(p.x))

print(log_likelihood_fit) # -9.946161673231583
```

Using some clever binary multiplication and Python comprehensions, we can consolidate that for loop and if expression into one line that returns the log_likelihood_fit. Similar to what we did in the maximum likelihood formula, we can use some binary subtraction between the true and false cases to mathematically eliminate one or the other. In this case, we multiply by 0 and therefore apply either the true or the false case, but not both, to the sum accordingly (Example 6-11).

Example 6-11. Consolidating our log likelihood logic into a single line

```
log_likelihood_fit = sum(log(logistic_function(p.x)) * p.y +
                         log(1.0 - logistic_function(p.x)) * (1.0 - p.y)
                         for p in patient_data)
```

If we were to express the likelihood of the fit in mathematic notation, this is what it would look like. Note that $f(x_i)$ is the logistic function for a given input variable x_i:

$$\text{log likelihood fit} = \sum_{i=1}^{n} (log(f(x_i)) \times y_i) + (log(1.0 - f(x_i)) \times (1 - y_i))$$

As calculated in Examples 6-10 and 6-11, we have -9.9461 as our log likelihood of the fit. We need one more datapoint to calculate the R^2: the log likelihood that estimates without using any input variables and simply uses the number of true cases divided by all cases (effectively leaving only the intercept). Note we can count the number of symptomatic cases by summing all the y-values together $\sum y_i$, because only the 1s and not the 0s will count into the sum. Here is the formula:

$$\text{log likelihood} = \sum \left(log\left(\frac{\sum y_i}{n}\right) \times y_i + log\left(1 - \frac{\sum y_i}{n}\right) \times (1 - y_i) \right)$$

Here is the expanded Python equivalent of this formula applied in Example 6-12.

Example 6-12. Log likelihood of patients

```
import pandas as pd
from math import log, exp

patient_data = list(pd.read_csv('https://bit.ly/33ebs2R', delimiter=",") \
    .itertuples())

likelihood = sum(p.y for p in patient_data) / len(patient_data)

log_likelihood = 0.0

for p in patient_data:
    if p.y == 1.0:
        log_likelihood += log(likelihood)
    elif p.y == 0.0:
        log_likelihood += log(1.0 - likelihood)

print(log_likelihood) # -14.341070198709906
```

To consolidate this logic and reflect the formula, we can compress that for loop and if expression into a single line, using some binary multiplication logic to handle both true and false cases (Example 6-13).

Example 6-13. Consolidating the log likelihood into a single line

```
log_likelihood = sum(log(likelihood)*p.y + log(1.0 - likelihood)*(1.0 - p.y) \
        for p in patient_data)
```

Finally, just plug these values in and get your R^2:

$$R^2 = \frac{(\text{log likelihood}) - (\text{log likelihood fit})}{(\text{log likelihood})}$$

$$R^2 = \frac{-14.341 - (-9.9461)}{-14.341}$$

$$R^2 = 0.306456$$

And here is the Python code shown in Example 6-14, calculating the R^2 in its entirety.

Example 6-14. Calculating the R^2 for a logistic regression

```
import pandas as pd
from math import log, exp

patient_data = list(pd.read_csv('https://bit.ly/33ebs2R', delimiter=",") \
                        .itertuples())

# Declare fitted logistic regression
b0 = -3.17576395
b1 = 0.69267212

def logistic_function(x):
    p = 1.0 / (1.0 + exp(-(b0 + b1 * x)))
    return p

# calculate the log likelihood of the fit
log_likelihood_fit = sum(log(logistic_function(p.x)) * p.y +
                         log(1.0 - logistic_function(p.x)) * (1.0 - p.y)
                         for p in patient_data)

# calculate the log likelihood without fit
likelihood = sum(p.y for p in patient_data) / len(patient_data)

log_likelihood = sum(log(likelihood) * p.y + log(1.0 - likelihood) * (1.0 - p.y) \
        for p in patient_data)

# calculate R-Square
r2 = (log_likelihood - log_likelihood_fit) / log_likelihood

print(r2)  # 0.306456105756576
```

OK, so we got an $R^2 = 0.306456$, so do hours of chemical exposure explain whether someone shows symptoms? As we learned in Chapter 5 on linear regression, a poor fit will be closer to an R^2 of 0.0 and a greater fit will be closer to 1.0. Therefore, we can conclude that hours of exposure is mediocre for predicting symptoms, as the R^2 is 0.30645. There must be variables other than time exposure that better predict if someone will show symptoms. This makes sense because we have a large mix of patients showing symptoms versus not showing symptoms for most of our observed data, as shown in Figure 6-12.

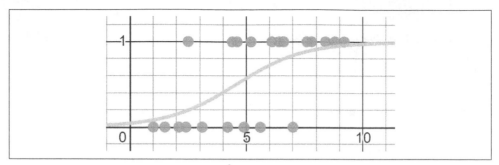

Figure 6-12. Our data has a mediocre R^2 of 0.30645 because there is a lot of variance in the middle of our curve

But if we did have a clean divide in our data, where 1 and 0 outcomes are cleanly separated as shown in Figure 6-13, we would have a perfect R^2 of 1.0.

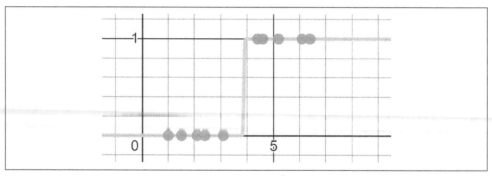

Figure 6-13. This logistic regression has a perfect R^2 of 1.0 because there is a clean divide in outcomes predicted by hours of exposure

P-Values

Just like linear regression, we are not done just because we have an R^2. We need to investigate how likely we would have seen this data by chance rather than because of an actual relationship. This means we need a p-value.

To do this, we will need to learn a new probability distribution called the *chi-square distribution*, annotated as χ^2 distribution. It is continuous and used in several areas of statistics, including this one!

If we take each value in a standard normal distribution (mean of 0 and standard deviation of 1) and square it, that will give us the χ^2 distribution with one degree of freedom. For our purposes, the degrees of freedom will depend on how many parameters n are in our logistic regression, which will be $n - 1$. You can see examples of different degrees of freedom in Figure 6-14.

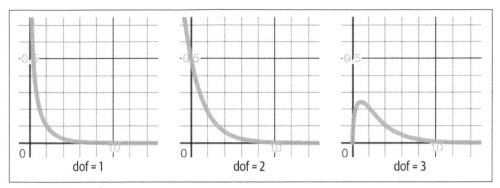

Figure 6-14. A χ^2 distribution with differing degrees of freedom

Since we have two parameters (hours of exposure and whether symptoms were shown), our degree of freedom will be 1 because $2 - 1 = 1$.

We will need the log likelihood fit and log likelihood as calculated in the previous subsection on R^2. Here is the formula that will produce the χ^2 value we need to look up:

$$\chi^2 = 2((\text{log likelihood fit}) - (\text{log likelihood}))$$

We then take that value and look up the probability from the χ^2 distribution. That will give us our p-value:

$$\text{p-value} = \text{chi}(2((\text{log likelihood fit}) - (\text{log likelihood})))$$

Example 6-15 shows our p-value for a given fitted logistic regression. We use SciPy's chi2 module to use the chi-square distribution.

Example 6-15. Calculating a p-value for a given logistic regression

```
import pandas as pd
from math import log, exp
from scipy.stats import chi2

patient_data = list(pd.read_csv('https://bit.ly/33ebs2R', delimiter=",").itertuples())

# Declare fitted logistic regression
b0 = -3.17576395
b1 = 0.69267212

def logistic_function(x):
    p = 1.0 / (1.0 + exp(-(b0 + b1 * x)))
    return p
```

```
# calculate the log likelihood of the fit
log_likelihood_fit = sum(log(logistic_function(p.x)) * p.y +
                         log(1.0 - logistic_function(p.x)) * (1.0 - p.y)
                         for p in patient_data)

# calculate the log likelihood without fit
likelihood = sum(p.y for p in patient_data) / len(patient_data)

log_likelihood = sum(log(likelihood) * p.y + log(1.0 - likelihood) * (1.0 - p.y) \
                     for p in patient_data)

# calculate p-value
chi2_input = 2 * (log_likelihood_fit - log_likelihood)
p_value = chi2.pdf(chi2_input, 1) # 1 degree of freedom (n - 1)

print(p_value)  # 0.0016604875618753787
```

So we have a p-value of 0.00166, and if our threshold for signifiance is .05, we say this data is statistically significant and was not by random chance.

Train/Test Splits

As covered in Chapter 5 on linear regression, we can use train/test splits as a way to validate machine learning algorithms. This is the more machine learning approach to assessing the performance of a logistic regression. While it is a good idea to rely on traditional statistical metrics like R^2 and p-values, when you are dealing with more variables, this becomes less practical. This is where train/test splits come in handy once again. To review, Figure 6-15 visualizes a three-fold cross-validation alternating a testing dataset.

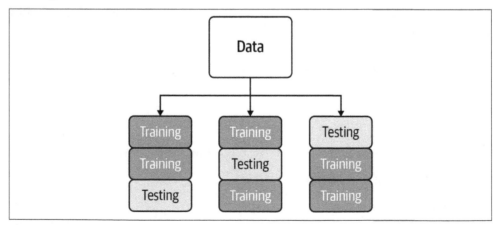

Figure 6-15. A three-fold cross-validation alternating each third of the dataset as a testing dataset

In Example 6-16 we perform a logistic regression on the employee-retention dataset, but we split the data into thirds. We then alternate each third as the testing data. Finally, we summarize the three accuracies with an average and standard deviation.

Example 6-16. Performing a logistic regression with three-fold cross-validation

```
import pandas as pd
from sklearn.linear_model import LogisticRegression
from sklearn.model_selection import KFold, cross_val_score

# Load the data
df = pd.read_csv("https://tinyurl.com/y6r7qjrp", delimiter=",")

X = df.values[:, :-1]
Y = df.values[:, -1]

# "random_state" is the random seed, which we fix to 7
kfold = KFold(n_splits=3, random_state=7, shuffle=True)
model = LogisticRegression(penalty='none')
results = cross_val_score(model, X, Y, cv=kfold)

print("Accuracy Mean: %.3f (stdev=%.3f)" % (results.mean(), results.std()))
```

We can also use random-fold validation, leave-one-out cross-validation, and all the other folding variants we performed in Chapter 5. With that out of the way, let's talk about why accuracy is a bad measure for classification.

Confusion Matrices

Suppose a model observed people with the name "Michael" quit their job. The reason why first and last names are captured as input variables is indeed questionable, as it is doubtful someone's name has any impact on whether they quit. However, to simplify the example, let's go with it. The model then predicts that any person named "Michael" will quit their job.

Now this is where accuracy falls apart. I have one hundred employees, including one named "Michael" and another named "Sam." Michael is wrongly predicted to quit, and it is Sam that ends up quitting. What's the accuracy of my model? It is 98% because there were only two wrong predictions out of one hundred employees as visualized in Figure 6-16.

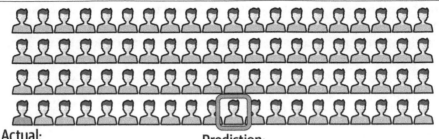

Actual:
This employee quits. Prediction was wrong, but I'm still 98% accurate!

Prediction:
This employee is named "Michael." This employee will quit.

Figure 6-16. The employee named "Michael" is predicted to quit, but it's actually another employee that does, giving us 98% accuracy

Especially for imbalanced data where the event of interest (e.g., a quitting employee) is rare, the accuracy metric is horrendously misleading for classification problems. If a vendor, consultant, or data scientist ever tries to sell you a classification system on claims of accuracy, ask for a confusion matrix.

A *confusion matrix* is a grid that breaks out the predictions against the actual outcomes showing the true positives, true negatives, false positives (type I error), and false negatives (type II error). Here is a confusion matrix presented in Figure 6-17.

	Actually quits (true)	Actually stays (false)
Predicted will quit (true)	0	1
Predicted will stay(false)	1	98

Figure 6-17. A simple confusion matrix

Generally, we want the diagonal values (top-left to bottom-right) to be higher because these reflect correct classifications. We want to evaluate how many employees who were predicted to quit actually did quit (true positives). Conversely, we also want to evaluate how many employees who were predicted to stay actually did stay (true negatives).

The other cells reflect wrong predictions, where an employee predicted to quit ended up staying (false positive), and where an employee predicted to stay ends up quitting (false negative).

What we need to do is dice up that accuracy metric into more specific accuracy metrics targeting different parts of the confusion matrix. Let's look at Figure 6-18, which adds some useful measures.

From the confusion matrix, we can derive all sorts of useful metrics beyond just accuracy. We can easily see that precision (how accurate positive predictions were) and sensitivity (rate of identified positives) are 0, meaning this machine learning model fails entirely at positive predictions.

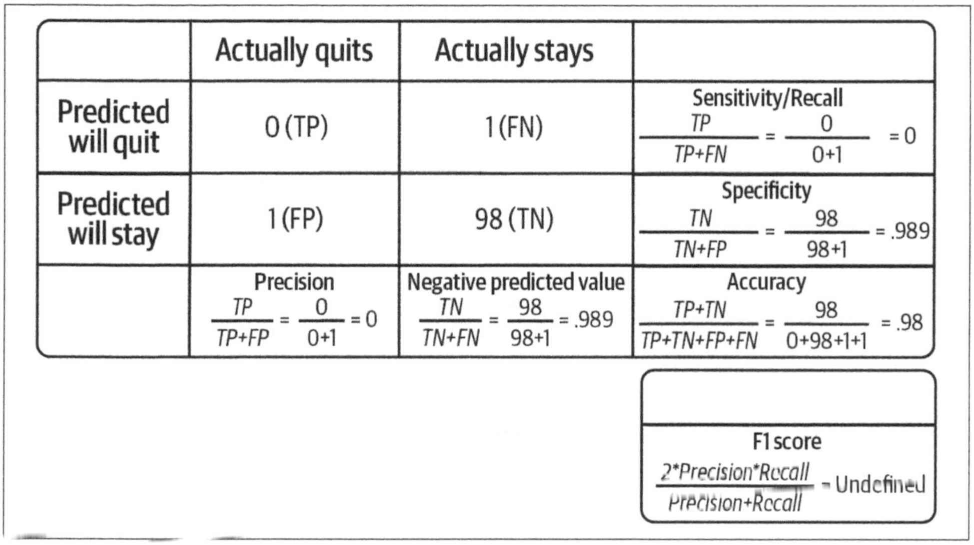

	Actually quits	Actually stays	
Predicted will quit	0 (TP)	1 (FN)	Sensitivity/Recall $\dfrac{TP}{TP+FN} = \dfrac{0}{0+1} = 0$
Predicted will stay	1 (FP)	98 (TN)	Specificity $\dfrac{TN}{TN+FP} = \dfrac{98}{98+1} = .989$
	Precision $\dfrac{TP}{TP+FP} = \dfrac{0}{0+1} = 0$	Negative predicted value $\dfrac{TN}{TN+FN} = \dfrac{98}{98+1} = .989$	Accuracy $\dfrac{TP+TN}{TP+TN+FP+FN} = \dfrac{98}{0+98+1+1} = .98$

$$\text{F1 score}$$
$$\dfrac{2*Precision*Recall}{Precision+Recall} = \text{Undefined}$$

Figure 6-18. Adding useful metrics to the confusion matrix

Example 6-17 shows how to use the confusion matrix API in SciPy on a logistic regression with a train/test split. Note that the confusion matrix is only applied to the testing dataset.

Example 6-17. Creating a confusion matrix for a testing dataset in SciPy

```
import pandas as pd
from sklearn.linear_model import LogisticRegression
from sklearn.metrics import confusion_matrix
from sklearn.model_selection import train_test_split

# Load the data
df = pd.read_csv('https://bit.ly/3cManTi', delimiter=",")

# Extract input variables (all rows, all columns but last column)
X = df.values[:, :-1]

# Extract output column (all rows, last column)\
Y = df.values[:, -1]
```

```
model = LogisticRegression(solver='liblinear')

X_train, X_test, Y_train, Y_test = train_test_split(X, Y, test_size=.33,
    random_state=10)
model.fit(X_train, Y_train)
prediction = model.predict(X_test)

"""
The confusion matrix evaluates accuracy within each category.
[[truepositives falsenegatives]
 [falsepositives truenegatives]]

The diagonal represents correct predictions,
so we want those to be higher
"""
matrix = confusion_matrix(y_true=Y_test, y_pred=prediction)
print(matrix)

"""
[[6 3]
 [4 5]]
"""
```

Bayes' Theorem and Classification

Do you recall Bayes' Theorem in Chapter 2? You can use Bayes' Theorem to bring in outside information to further validate findings on a confusion matrix. Figure 6-19 shows a confusion matrix of one thousand patients tested for a disease.

	Tests positive	Tests negative
At risk	198	2
Not at risk	50	750

Figure 6-19. A confusion matrix for a medical test identifying a disease

We are told that for patients that have a health risk, 99% will be identified successfully (sensitivity). Using the confusion matrix, we can see this mathematically checks out:

$$\text{sensitivity} = \frac{198}{198 + 2} = .99$$

But what if we flip the condition? What percentage of those who tested positive have the health risk (precision)? While we are flipping a conditional probability, we do not have to use Bayes' Theorem here because the confusion matrix gives us all the numbers we need:

$$\text{precision} = \frac{198}{198 + 50} = .798$$

OK, so 79.8% is not terrible, and that's the percentage of people who tested positive that actually have the disease. But ask yourself this…what are we assuming about our data? Is it representative of the population?

Some quick research found 1% of the population actually has the disease. There is an opportunity to use Bayes' Theorem here. We can account for the proportion of the population that actually has the disease and incorporate that into our confusion matrix findings. We then discover something significant.

$$P(\text{At Risk if Positive}) = \frac{P(\text{Positive if At Risk}) \times P(\text{At Risk})}{P(\text{Positive})}$$

$$P(\text{At Risk if Positive}) = \frac{.99 \times .01}{.248}$$

$$P(\text{At Risk if Positive}) = .0339$$

When we account for the fact that only 1% of the population is at risk, and 20% of our test patients are at risk, the probability of being at risk given a positive test is 3.39%! How did it drop from 99%? This just shows how easily we can get duped by probabilities that are high only in a specific sample like the vendor's one thousand test patients. So if this test has only a 3.39% probability of successfully identifying a true positive, we probably should not use it.

Receiver Operator Characteristics/Area Under Curve

When we are evaluating different machine learning configurations, we may end up with dozens, hundreds, or thousands of confusion matrices. These can be tedious to review, so we can summarize all of them with a *receiver operator characteristic (ROC) curve* as shown in Figure 6-20. This allows us to see each testing instance (each represented by a black dot) and find an agreeable balance between true positives and false positives.

We can also compare different machine learning models by creating separate ROC curves for each. For example, if in Figure 6-21 our top curve represents a logistic regression and the bottom curve represents a decision tree (a machine learning technique we did not cover in this book), we can see the performance of them side by side. The *area under the curve (AUC)* is a good metric for choosing which model to use. Since the top curve (logistic regression) has a greater area, this suggests it is a superior model.

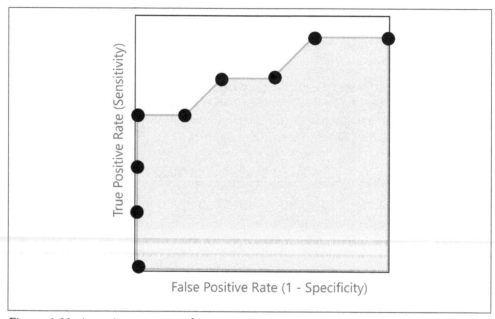

Figure 6-20. A receiver operator characteristic curve

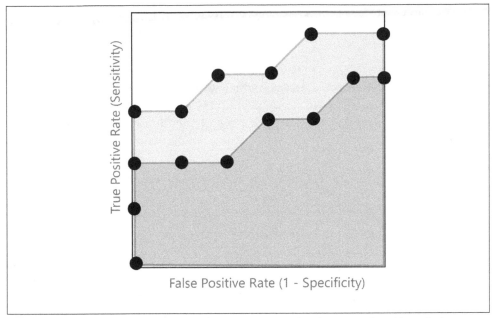

Figure 6-21. Comparing two models by their area under the curve (AUC) with their respective ROC curves

To use the AUC as a scoring metric, change the scoring parameter in the scikit learn API to use `roc_auc` as shown for a cross-validation in Example 6-18.

Example 6-18. Using the AUC as the scikit-learn parameter

```
# put Scikit_learn model here

results = cross_val_score(model, X, Y, cv=kfold, scoring='roc_auc')
print("AUC: %.3f (%.3f)" % (results.mean(), results.std()))
# AUC: 0.791 (0.051)
```

Class Imbalance

There is one last thing to cover before we close this chapter. As we saw earlier when discussing confusion matrices, *class imbalance*, which happens when data is not equally represented across every outcome class, is a problem in machine learning. Unfortunately, many problems of interest are imbalanced, such as disease prediction, security breaches, fraud detection, and so on. Class imbalance is still an open problem with no great solution. However, there are a few techniques you can try.

First, you can do obvious things like collect more data or try different models as well as use confusion matrices. All of this will help track poor predictions and proactively catch errors.

Another common technique is to duplicate samples in the minority class until it is equally represented in the dataset. Another technique is shown in Example 6-19 when doing your train-test splits. Pass the `stratify` option with the column containing the class values, and it will keep the class distribution consistent between the train-test split.

Example 6-19. Using the `stratify` option in scikit-learn to balance classes in the data

```
X, Y = ...
X_train, X_test, Y_train, Y_test =  \
        train_test_split(X, Y, test_size=.33, stratify=Y)
```

There is also a family of algorithms called SMOTE, which generate synthetic samples of the minority class. What would be most ideal though is to tackle the problem in a way that uses anomaly-detection models, which are deliberately designed for seeking out a rare event. These seek outliers, however, and are not necessarily a classification since they are unsupervised algorithms. All these techniques are beyond the scope of this book but are worth mentioning as they *might* provide better solutions to a given problem.

Conclusion

Logistic regression is the workhorse model for predicting probabilities and classifications on data. Logistic regressions can predict more than one category rather than just a true/false. You just build separate logistic regressions modeling whether or not it belongs to that category, and the model that produces the highest probability is the one that wins. You may discover that scikit-learn, for the most part, will do this for you and detect when your data has more than two classes.

In this chapter, we covered not just how to fit a logistic regression using gradient descent and scikit-learn but also statistical and machine learning approaches to validation. On the statistical front we covered the R^2 and p-value, and in machine learning we explored train/test splits, confusion matrices, and ROC/AUC.

If you want to learn more about logistic regression, probably the best resource to jump-start further is Josh Starmer's StatQuest playlist on Logistic Regression. I have to credit Josh's work in assisting some portions of this chapter, particularly in how to calculate R^2 and p-values for logistic regression. If nothing else, watch his videos for the fantastic opening jingles (*https://oreil.ly/tueJJ*)!

As always, you will find yourself walking between the two worlds of statistics and machine learning. Many books and resources right now cover logistic regression from a machine learning perspective, but try to seek out statistics resources too. There are advantages and disadvantages to both schools of thought, and you can win only by being adaptable to both!

Exercises

A dataset of three input variables RED, GREEN, and BLUE as well as an output variable LIGHT_OR_DARK_FONT_IND is provided here (*https://bit.ly/3imidqa*). It will be used to predict whether a light/dark font (0/1 respectively) will work for a given background color (specified by RGB values).

1. Perform a logistic regression on the preceding data, using three-fold cross-validation and accuracy as your metric.

2. Produce a confusion matrix comparing the predictions and actual data.

3. Pick a few different background colors (you can use an RGB tool like this one (*https://bit.ly/3FHywrZ*)) and see if the logistic regression sensibly chooses a light (0) or dark (1) font for each one.

4. Based on the preceding exercises, do you think logistic regression is effective for predicting a light or dark font for a given background color?

Answers are in Appendix B.

## CHAPTER 7

Neural Networks

A regression and classification technique that has enjoyed a renaissance over the past 10 years is neural networks. In the simplest definition, a *neural network* is a multilayered regression containing layers of weights, biases, and nonlinear functions that reside between input variables and output variables. *Deep learning* is a popular variant of neural networks that utilizes multiple "hidden" (or middle) layers of nodes containing weights and biases. Each node resembles a linear function before being passed to a nonlinear function (called an activation function). Just like linear regression, which we learned about in Chapter 5, optimization techniques like stochastic gradient descent are used to find the optimal weight and bias values to minimize the residuals.

Neural networks offer exciting solutions to problems previously difficult for computers to solve. From identifying objects in images to processing words in audio, neural networks have created tools that affect our everyday lives. This includes virtual assistants and search engines, as well as photo tools in our iPhones.

Given the media hoopla and bold claims dominating news headlines about neural networks, it may be surprising that they have been around since the 1950s. The reason for their sudden popularity after 2010 is due to the growing availability of data and computing power. The ImageNet challenge between 2011 and 2015 was probably the largest driver of the renaissance, boosting performance on classifying one thousand categories on 1.4 million images to an accuracy of 96.4%.

However, like any machine learning technique it only works on narrowly defined problems. Even projects to create "self-driving" cars do not use end-to-end deep learning, and primarily use hand-coded rule systems with convoluted neural networks acting as a "label maker" for identifying objects on the road. We will discuss this later in this chapter to understand where neural networks are actually used. But first we will build a simple neural network in NumPy, and then use scikit-learn as a library implementation.

When to Use Neural Networks and Deep Learning

Neural networks and deep learning can be used for classification and regression, so how do they size up to linear regression, logistic regression, and other types of machine learning? You might have heard the expression "when all you have is a hammer, everything starts to look like a nail." There are advantages and disadvantages that are situational for each type of algorithm. Linear regression and logistic regression, as well as gradient boosted trees (which we did not cover in this book), do a pretty fantastic job making predictions on structured data. Think of structured data as data that is easily represented as a table, with rows and columns. But perceptual problems like image classification are much less structured, as we are trying to find fuzzy correlations between groups of pixels to identify shapes and patterns, not rows of data in a table. Trying to predict the next four or five words in a sentence being typed, or deciphering the words being said in an audio clip, are also perceptual problems and examples of neural networks being used for natural language processing.

In this chapter, we will primarily focus on simple neural networks with only one hidden layer.

Variants of Neural Networks

Variants of neural networks include convolutional neural networks, which are often used for image recognition. Long short-term memory (LSTM) is used for predicting time series, or forecasting. Recurrent neural networks are often used for text-to-speech applications.

Is Using a Neural Network Overkill?

Using neural networks for the upcoming example is probably overkill, as a logistic regression would probably be more practical. Even a formulaic approach can be used (*https://oreil.ly/M4W8i*). However, I have always been a fan of understanding complex techniques by applying them to simple problems. You learn about the strengths and limitations of the technique rather than be distracted by large datasets. So with that in mind, try not to use neural networks where simpler models will be more practical. We will break this rule in this chapter for the sake of understanding the technique.

A Simple Neural Network

Here is a simple example to get a feel for neural networks. I want to predict whether a font should be light (1) or dark (0) for a given color background. Here are a few examples of different background colors in Figure 7-1. The top row looks best with light font, and the bottom row looks best with dark font.

Figure 7-1. Light background colors look best with dark font, and dark background colors look best with light font

In computer science one way to represent a color is with RGB values, or the red, green, and blue values. Each of these values is between 0 and 255 and expresses how these three colors are mixed to create the desired color. For example, if we express the RGB as (red, green, blue), then dark orange would have an RGB of (255,140,0) and pink would be (255,192,203). Black would be (0,0,0) and white would be (255,255,255).

From a machine learning and regression perspective, we have three numeric input variables red, green, and blue to capture a given background color. We need to fit a function to these input variables and output whether a light (1) or dark (0) font should be used for that background color.

Representing Colors Through RGB

There are hundreds of color picker palettes online to experiment with RGB values. W3 Schools has one here (*https://oreil.ly/T57gu*).

Note this example is not far from how neural networks work recognizing images, as each pixel is often modeled as three numeric RGB values. In this case, we are just focusing on one "pixel" as a background color.

Let's start high level and put all the implementation details aside. We are going to approach this topic like an onion, starting with a higher understanding and peeling away slowly into the details. For now, this is why we simply label as "mystery math" a process that takes inputs and produces outputs. We have three numeric input variables R, G, and B, which are processed by this mystery math. Then it outputs a prediction between 0 and 1 as shown in Figure 7-2.

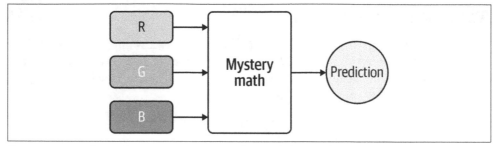

Figure 7-2. We have three numeric RGB values used to make a prediction for a light or dark font

This prediction output expresses a probability. Outputting probabilities is the most common model for classification with neural networks. Once we replace RGB with their numerical values, we see that less than 0.5 will suggest a dark font whereas greater than 0.5 will suggest a light font as demonstrated in Figure 7-3.

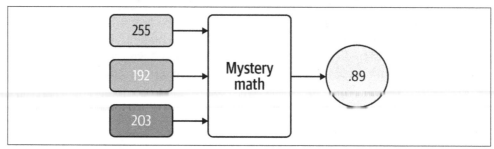

Figure 7-3. If we input a background color of pink (255,192,203), then the mystery math recommends a light font because the output probability 0.89 is greater than 0.5

So what is going on inside that mystery math black box? Let's take a look in Figure 7-4.

We are missing another piece of this neural network, the activation functions, but we will get to that shortly. Let's first understand what's going on here. The first layer on the left is simply an input of the three variables, which in this case are the red, green, and blue values. In the hidden (middle) layer, notice that we produce three *nodes*, or functions of weights and biases, between the inputs and outputs. Each node essentially is a linear function with slopes W_i and intercepts B_i being multiplied and summed with input variables X_i. There is a weight W_i between each input node and hidden node, and another set of weights between each hidden node and output node. Each hidden and output node gets an additional bias B_i added.

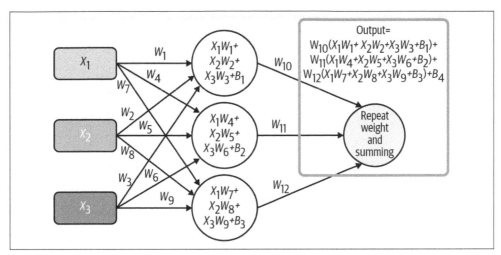

Figure 7-4. The hidden layer of the neural network applies weight and bias values to each input variable, and the output layer applies another set of weights and biases to that output

Notice the output node repeats the same operation, taking the resulting weighted and summed outputs from the hidden layer and making them inputs into the final layer, where another set of weights and biases will be applied.

In a nutshell, this is a regression just like linear or logistic regression, but with many more parameters to solve for. The weight and bias values are analogous to the m and b, or β_1 and β_0, parameters in a linear regression. We do use stochastic gradient descent and minimize loss just like linear regression, but we need an additional tool called backpropagation to untangle the weight W_i and bias B_i values and calculate their partial derivatives using the chain rule. We will get to that later in this chapter, but for now let's assume we have the weight and bias values optimized. We need to talk about activation functions first.

Activation Functions

Let's bring in the activation functions next. An *activation function* is a nonlinear function that transforms or compresses the weighted and summed values in a node, helping the neural network separate the data effectively so it can be classified. Let's take a look at Figure 7-5. If you do not have the activation functions, your hidden layers will not be productive and will perform no better than a linear regression.

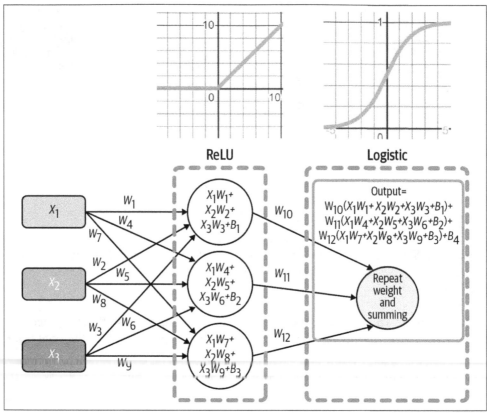

Figure 7-5. Applying activation functions

The *ReLU activation function* will zero out any negative outputs from the hidden nodes. If the weights, biases, and inputs multiply and sum to a negative number, it will be converted to 0. Otherwise the output is left alone. Here is the graph for ReLU (Figure 7-6) using SymPy (Example 7-1).

Example 7-1. Plotting the ReLU function

```
from sympy import *

# plot relu
x = symbols('x')
relu = Max(0, x)
plot(relu)
```

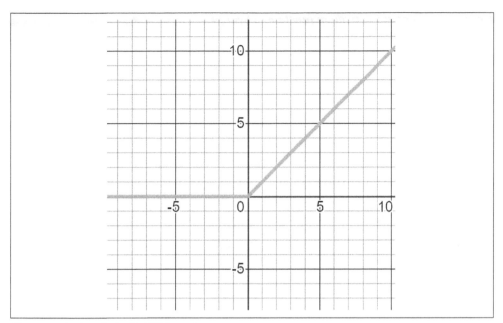

Figure 7-6. Graph for ReLU function

ReLU is short for "rectified linear unit," but that is just a fancy way of saying "turn negative values into 0." ReLU has gotten popular for hidden layers in neural networks and deep learning because of its speed and mitigation of the vanishing gradient problem (*https://oreil.ly/QGlM7*). Vanishing gradients occur when the partial derivative slopes get so small they prematurely approach 0 and bring training to a screeching halt.

The output layer has an important job: it takes the piles of math from the hidden layers of the neural network and turns them into an interpretable result, such as presenting classification predictions. The output layer for this particular neural network uses the *logistic activation function*, which is a simple sigmoid curve. If you read Chapter 6, the logistic (or sigmoid) function should be familiar, and it demonstrates that logistic regression is acting as a layer in our neural network. The output node weights, biases, and sums each of the incoming values from the hidden layer. After that, it passes the resulting value through the logistic function so it outputs a number between 0 and 1. Much like logistic regression in Chapter 6, this represents a probability that the given color input into the neural network recommends a light font. If it is greater than or equal to 0.5, the neural network is suggesting a light font, but less than that will advise a dark font.

Here is the graph for the logistic function (Figure 7-7) using SymPy (Example 7-2).

Example 7-2. Logistic activation function in SymPy

```
from sympy import *

# plot logistic
x = symbols('x')
logistic = 1 / (1 + exp(-x))
plot(logistic)
```

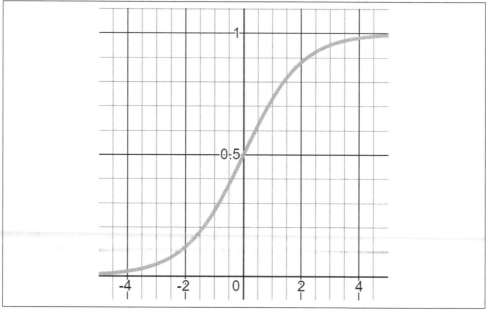

Figure 7-7. Logistic activation function

Note that when we pass a node's weighted, biased, and summed value through an activation function, we now call that an *activated output*, meaning it has been filtered through the activation function. When the activated output leaves the hidden layer, the signal is ready to be fed into the next layer. The activation function could have strengthened, weakened, or left the signal as is. This is where the brain and synapse metaphor for neural networks comes from.

Given the potential for complexity, you might be wondering if there are other activation functions. Some common ones are shown in Table 7-1.

Table 7-1. Common activation functions

Name	Typical layer used	Description	Notes
Linear	Output	Leaves values as is	Not commonly used
Logistic	Output	S-shaped sigmoid curve	Compresses values between 0 and 1, often assists binary classification
Tangent Hyperbolic	Hidden	tanh, S-shaped sigmoid curve between -1 and 1	Assists in "centering" data by bringing mean close to 0
ReLU	Hidden	Turns negative values to 0	Popular activation faster than sigmoid and tanh, mitigates vanishing gradient problems and computationally cheap
Leaky ReLU	Hidden	Multiplies negative values by 0.01	Controversial variant of ReLU that marginalizes rather than eliminates negative values
Softmax	Output	Ensures all output nodes add up to 1.0	Useful for multiple classifications and rescaling outputs so they add up to 1.0

This is not a comprehensive list of activation functions, and in theory any function could be an activation function in a neural network.

While this neural network seemingly supports two classes (light or dark font), it actually is modeled to one class: whether or not a font should be light (1) or not (0). If you wanted to support multiple classes, you could add more output nodes for each class. For instance, if you are trying to recognize handwritten digits 0–9, there would be 10 output nodes representing the probability a given image is each of those numbers. You might consider using softmax as the output activation when you have multiple classes as well. Figure 7-8 shows an example of taking a pixellated image of a digit, where the pixels are broken up as individual neural network inputs and then passed through two middle layers, and then an output layer with 10 nodes representing probabilities for 10 classes (for the digits 0–9).

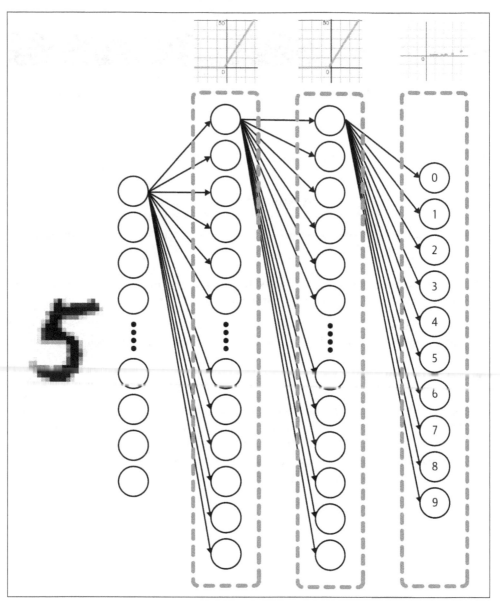

Figure 7-8. A neural network that takes each pixel as an input and predicts what digit the image contains

An example of using the MNIST dataset on a neural network can be found in Appendix A.

I Don't Know What Activation Function to Use!

If you are unsure what activations to use, current best practices gravitate toward ReLU for middle layers and logistic (sigmoid) for output layer. If you have multiple classifications in the output, use softmax for the output layer.

Forward Propagation

Let's capture what we have learned so far using NumPy. Note I have not optimized the parameters (our weight and bias values) yet. We are going to initialize those with random values.

Example 7-3 is the Python code to create a simple feed-forward neural network that is not optimized yet. *Feed forward* means we are simply inputting a color into the neural network and seeing what it outputs. The weights and biases are randomly initialized and will be optimized later in this chapter, so do not expect a useful output yet.

Example 7-3. A simple forward propagation network with random weight and bias values

```
import numpy as np
import pandas as pd
from sklearn.model_selection import train_test_split

all_data = pd.read_csv("https://tinyurl.com/y2qmhfsr")

# Extract the input columns, scale down by 255
all_inputs = (all_data.iloc[:, 0:3].values / 255.0)
all_outputs = all_data.iloc[:, -1].values

# Split train and test data sets
X_train, X_test, Y_train, Y_test = train_test_split(all_inputs, all_outputs,
    test_size=1/3)
n = X_train.shape[0] # number of training records

# Build neural network with weights and biases
# with random initialization
w_hidden = np.random.rand(3, 3)
w_output = np.random.rand(1, 3)

b_hidden = np.random.rand(3, 1)
b_output = np.random.rand(1, 1)

# Activation functions
relu = lambda x: np.maximum(x, 0)
logistic = lambda x: 1 / (1 + np.exp(-x))
```

```
# Runs inputs through the neural network to get predicted outputs
def forward_prop(X):
    Z1 = w_hidden @ X + b_hidden
    A1 = relu(Z1)
    Z2 = w_output @ A1 + b_output
    A2 = logistic(Z2)
    return Z1, A1, Z2, A2

# Calculate accuracy
test_predictions = forward_prop(X_test.transpose())[3] # grab only output layer, A2
test_comparisons = np.equal((test_predictions >= .5).flatten().astype(int), Y_test)
accuracy = sum(test_comparisons.astype(int) / X_test.shape[0])
print("ACCURACY: ", accuracy)
```

A couple of things to note here. The dataset containing the RGB input values as well as output value (1 for light and 0 for dark) are contained in this CSV file (*https://oreil.ly/1TZIK*). I am scaling down the input columns R, G, and B values by a factor of 1/255 so they are between 0 and 1. This will help the training later so the number space is compressed.

Note I also separated 2/3 of the data for training and 1/3 for testing using scikit-learn, which we learned how to do in Chapter 5. n is simply the number of training data records.

Now bring your attention to the lines of code shown in Example 7-4.

Example 7-4. The weight matrices and bias vectors in NumPy

```
# Build neural network with weights and biases
# with random initialization
w_hidden = np.random.rand(3, 3)
w_output = np.random.rand(1, 3)

b_hidden = np.random.rand(3, 1)
b_output = np.random.rand(1, 1)
```

These are declaring our weights and biases for both the hidden and output layers of our neural network. This may not be obvious yet but matrix multiplication is going to make our code powerfully simple using linear algebra and NumPy.

The weights and biases are going to be initialized as random values between 0 and 1. Let's look at the weight matrices first. When I ran the code I got these matrices:

$$W_{hidden} = \begin{bmatrix} 0.034535 & 0.5185636 & 0.81485028 \\ 0.3329199 & 0.53873853 & 0.96359003 \\ 0.19808306 & 0.45422182 & 0.36618893 \end{bmatrix}$$

$$W_{output} = \begin{bmatrix} 0.82652072 & 0.30781539 & 0.93095565 \end{bmatrix}$$

Note that W_{hidden} are the weights in the hidden layer. The first row represents the first node weights W_1, W_2, and W_3. The second row is the second node with weights W_4, W_5, and W_6. The third row is the third node with weights W_7, W_8, and W_9.

The output layer has only one node, meaning its matrix has only one row with weights W_{10}, W_{11}, and W_{12}.

See a pattern here? Each node is represented as a row in a matrix. If there are three nodes, there are three rows. If there is one node, there is one row. Each column holds a weight value for that node.

Let's look at the biases, too. Since there is one bias per node, there are going to be three rows of biases for the hidden layer and one row of biases for the output layer. There's only one bias per node so there will be only one column:

$$B_{hidden} = \begin{bmatrix} 0.41379442 \\ 0.81666079 \\ 0.07511252 \end{bmatrix}$$

$$B_{output} = \begin{bmatrix} 0.58018555 \end{bmatrix}$$

Now let's compare these matrix values to our visualized neural network as shown in Figure 7-9.

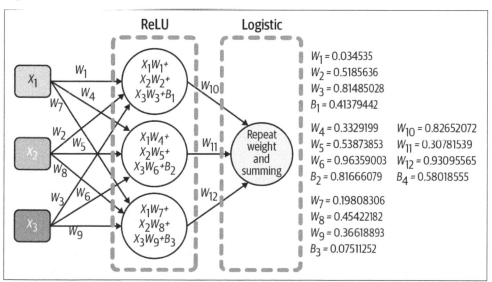

Figure 7-9. Visualizing our neural network against the weight and bias matrix values

So besides being esoterically compact, what is the benefit of these weights and biases in this matrix form? Let's bring our attention to these lines of code in Example 7-5.

Example 7-5. The activation functions and forward propagation function for our neural network

```
# Activation functions
relu = lambda x: np.maximum(x, 0)
logistic = lambda x: 1 / (1 + np.exp(-x))

# Runs inputs through the neural network to get predicted outputs
def forward_prop(X):
    Z1 = w_hidden @ X + b_hidden
    A1 = relu(Z1)
    Z2 = w_output @ A1 + b_output
    A2 = logistic(Z2)
    return Z1, A1, Z2, A2
```

This code is important because it concisely executes our entire neural network using matrix multiplication and matrix-vector multiplication. We learned about these operations in Chapter 4. It runs a color of three RGB inputs through the weights, biases, and activation functions in just a few lines of code.

I first declare the relu() and logistic() activation functions, which literally take a given input value and return the output value from the curve. The forward_prop() function executes our entire neural network for a given color input X containing the R, G, and B values. It returns the matrix outputs from four stages: Z1, A1, Z2, and A2. The "1" and "2" indicate the operations belong to layers 1 and 2 respectively. The "Z" indicates an unactivated output from the layer, and "A" is activated output from the layer.

The hidden layer is represented by Z1 and A1. Z1 is the weights and biases applied to X. Then A1 takes that output from Z1 and pushes it through the activation ReLU function. Z2 takes the output from A1 and applies the output layer weights and biases. That output is in turn pushed through the activation function, the logistic curve, and becomes A2. The final stage, A2, is the prediction probability from the output layer, returning a value between 0 and 1. We call it A2 because it is the "activated" output from layer 2.

Let's break this down in more detail starting with Z1:

$$Z_1 = W_{hidden}X + B_{hidden}$$

First we perform matrix-vector multiplication between W_{hidden} and the input color X. We multiply each row of W_{hidden} (each row being a set of weights for a node) with the vector X (the RGB color input values). We then add the biases to that result, as shown in Figure 7-10.

$$Z_1 = W_{hidden}X + B_{hidden}$$

$$Z_1 = \begin{bmatrix} 0.034535 & 0.5185636 & 0.81485028 \\ 0.3329199 & 0.53873853 & 0.96359003 \\ 0.19808306 & 0.45422182 & 0.36618893 \end{bmatrix} \begin{bmatrix} 0.82652072 \\ 0.30781539 \\ 0.93095565 \end{bmatrix} + \begin{bmatrix} 0.41379442 \\ 0.81666079 \\ 0.07511252 \end{bmatrix}$$

$$Z_1 = \begin{bmatrix} 0.946755221909086 \\ 1.33805678888247 \\ 0.644441873391768 \end{bmatrix} + \begin{bmatrix} 0.41379442 \\ 0.81666079 \\ 0.07511252 \end{bmatrix}$$

$$Z_1 = \begin{bmatrix} 1.36054964190909 \\ 2.15471757888247 \\ 0.719554393391768 \end{bmatrix}$$

Figure 7-10. Applying the hidden layer weights and biases to an input X using matrix-vector multiplication as well as vector addition

That Z_1 vector is the raw output from the hidden layer, but we still need to pass it through the activation function to turn Z_1 into A_1. Easy enough. Just pass each value in that vector through the ReLU function and it will give us A_1. Because all the values are positive, it should not have an impact.

$$A_1 = ReLU(Z_1)$$

$$A_1 = \begin{bmatrix} ReLU(1.36054964190909) \\ ReLU(2.15471757888247) \\ ReLU(0.719554393391768) \end{bmatrix} = \begin{bmatrix} 1.36054964190909 \\ 2.15471757888247 \\ 0.719554393391768 \end{bmatrix}$$

Now let's take that hidden layer output A_1 and pass it through the final layer to get Z_2 and then A_2. A_1 becomes the input into the output layer.

$$Z_2 = W_{output}A_1 + B_{output}$$

$$Z_2 = [0.82652072 \ 0.3078159 \ 0.93095565] \begin{bmatrix} 1.36054964190909 \\ 2.15471757888247 \\ 0.719554393391768 \end{bmatrix} + [0.58018555]$$

$$Z_2 = [2.45765202842636] + [0.58018555]$$

$$Z_2 = [3.03783757842636]$$

Finally, pass this single value in Z_2 through the activation function to get A_2. This will produce a prediction of approximately 0.95425:

$$A_2 = logistic(Z_2)$$

$$A_2 = logistic([3.0378364795204])$$

$$A_2 = 0.954254478103241$$

That executes our entire neural network, although we have not trained it yet. But take a moment to appreciate that we have taken all these input values, weights, biases, and nonlinear functions and turned them all into a single value that will provide a prediction.

Again, A2 is the final output that makes a prediction whether that background color need a light (1) or dark (0) font. Even though our weights and biases have not been optimized yet, let's calculate our accuracy as shown in Example 7-6. Take the testing dataset X_test, transpose it, and pass it through the forward_prop() function but only grab the A2 vector with the predictions for each test color. Then compare the predictions to the actuals and calculate the percentage of correct predictions.

Example 7-6. Calculating accuracy

```
# Calculate accuracy
test_predictions = forward_prop(X_test.transpose())[3]  # grab only A2
test_comparisons = np.equal((test_predictions >= .5).flatten().astype(int), Y_test)
accuracy = sum(test_comparisons.astype(int) / X_test.shape[0])
print("ACCURACY: ", accuracy)
```

When I run the whole code in Example 7-3, I roughly get anywhere from 55% to 67% accuracy. Remember, the weights and biases are randomly generated so answers will vary. While this may seem high given the parameters were randomly generated, remember that the output predictions are binary: light or dark. Therefore, a random coin flip might as well produce this outcome for each prediction, so this number should not be surprising.

 Do Not Forget to Check for Imbalanced Data!

As discussed in Chapter 6, do not forget to analyze your data to check for imbalanced classes. This whole background color dataset is a little imbalanced: 512 colors have output of 0 and 833 have an output of 1. This can skew accuracy and might be why our random weights and biases gravitate higher than 50% accuracy. If the data is extremely imbalanced (as in 99% of the data is one class), then remember to use confusion matrices to track the false positives and false negatives.

Does everything structurally make sense so far? Feel free to review everything up to this point before moving on. We just have one final piece to cover: optimizing the weights and biases. Hit the espresso machine or nitro coffee bar, because this is the most involved math we will be doing in this book!

Backpropagation

Before we start using stochastic gradient descent to optimize our neural network, a challenge we have is figuring out how to change each of the weight and bias values accordingly, even though they all are tangled together to create the output variable, which then is used to calculate the residuals. How do we find the derivative of each weight W_i and bias B_i variable? We need to use the chain rule, which we covered in Chapter 1.

Calculating the Weight and Bias Derivatives

We are not quite ready to apply stochastic gradient descent to train our neural network. We have to get the partial derivatives with respect to the weights W_i and biases B_i, and we have the chain rule to help us.

While the process is largely the same, there is a complication using stochastic gradient descent on neural networks. The nodes in one layer feed their weights and biases into the next layer, which then applies another set of weights and biases. This creates an onion-like nesting we need to untangle, starting with the output layer.

During gradient descent, we need to figure out which weights and biases should be adjusted, and by how much, to reduce the overall cost function. The cost for a single prediction is going to be the squared output of the neural network A_2 minus the actual value Y:

$$C = (A_2 - Y)^2$$

But let's peel back a layer. That activated output A_2 is just Z_2 with the activation function:

$$A_2 = sigmoid(Z_2)$$

Z_2 in turn is the output weights and biases applied to activation output A_1, which comes from the hidden layer:

$$Z_2 = W_2 A_1 + B_2$$

A_1 is built off Z_1 which is passed through the ReLU activation function:

$$A_1 = ReLU(Z_1)$$

Finally, Z_1 is the input x-values weighted and biased by the hidden layer:

$$Z_1 = W_1 X + B_1$$

We need to find the weights and biases contained in the W_1, B_1, W_2, and B_2 matrices and vectors that will minimize our loss. By nudging their slopes, we can change the weights and biases that have the most impact in minimizing loss. However, each little nudge on a weight or bias is going to propagate all the way to the loss function on the outer layer. This is where the chain rule can help us figure out this impact.

Let's focus on finding the relationship on a weight (W_2) from the output layer and the cost function C. A change in the weight W_2 results in a change to the unactivated output Z_2. That then changes the activated output A_2, which changes the cost function C. Using the chain rule, we can define the derivative of C with respect to W_2 as this:

$$\frac{dC}{dW_2} = \frac{dZ_2}{dW_2}\frac{dA_2}{dZ_2}\frac{dC}{dA_2}$$

When we multiply these three gradients together, we get a measure of how much a change to W_2 will change the cost function C.

Now we will calculate these three derivatives. Let's use SymPy to calculate the derivative of the cost function with respect to A_2 in Example 7-7.

$$\frac{dC}{dA_2} = 2A_2 - 2y$$

Example 7-7. Calculating the derivative of the cost function with respect to A_2

```
from sympy import *

A2, y = symbols('A2 Y')
C = (A2 - Y)**2
dC_dA2 = diff(C, A2)
print(dC_dA2) # 2*A2 - 2*Y
```

Next, let's get the derivative of A_2 with respect to Z_2 (Example 7-8). Remember that A_2 is the output of an activation function, in this case the logistic function. So we really are just taking the derivative of a sigmoid curve.

$$\frac{dA_2}{dZ_2} = \frac{e^{-Z_2}}{\left(1 + e^{-Z_2}\right)^2}$$

Example 7-8. Finding the derivative of A_2 with respect to Z_2

```
from sympy import *

Z2 = symbols('Z2')

logistic = lambda x: 1 / (1 + exp(-x))

A2 = logistic(Z2)
dA2_dZ2 = diff(A2, Z2)
print(dA2_dZ2) # exp(-Z2)/(1 + exp(-Z2))**2
```

The derivative of Z_2 with respect to W_2 is going to work out to be A_1, as it is just a linear function and going to return the slope (Example 7-9).

$$\frac{dZ_2}{dW_2} - A_1$$

Example 7-9. Derivative of Z_2 with respect to W_2

```
from sympy import *

A1, W2, B2 = symbols('A1, W2, B2')

Z2 = A1*W2 + B2
dZ2_dW2 = diff(Z2, W2)
print(dZ2_dW2) # A1
```

Putting it all together, here is the derivative to find how much a change in a weight in W_2 affects the cost function C:

$$\frac{dC}{dw_2} = \frac{dZ_2}{dw_2}\frac{dA_2}{dZ_2}\frac{dC}{dA_2} = (A_1)\left(\frac{e^{-Z_2}}{\left(1 + e^{-Z_2}\right)^2}\right)(2A_2 - 2y)$$

When we run an input X with the three input R, G, and B values, we will have values for A_1, A_2, Z_2, and y.

Don't Get Lost in the Math!

It is easy to get lost in the math at this point and forget what you were trying to achieve in the first place, which is finding the derivative of the cost function with respect to a weight (W_2) in the output layer. When you find yourself in the weeds and forgetting what you were trying to do, then step back, go for a walk, get a coffee, and remind yourself what you were trying to accomplish. If you cannot, you should start over from the beginning and work your way to the point you got lost.

However, this is just one component of the neural network, the derivative for W_2. Here are the SymPy calculations in Example 7-10 for the rest of the partial derivatives we will need for chaining.

Example 7-10. Calculating all the partial derivatives we will need for our neural network

```
from sympy import *

W1, W2, B1, B2, A1, A2, Z1, Z2, X, Y = \
    symbols('W1 W2 B1 B2 A1 A2 Z1 Z2 X Y')

# Calculate derivative of cost function with respect to A2
C = (A2 - Y)**2
dC_dA2 = diff(C, A2)
print("dC_dA2 = ", dC_dA2) # 2*A2 - 2*Y

# Calculate derivative of A2 with respect to Z2
logistic = lambda x: 1 / (1 + exp(-x))
_A2 = logistic(Z2)
dA2_dZ2 = diff(_A2, Z2)
print("dA2_dZ2 = ", dA2_dZ2) # exp(-Z2)/(1 + exp(-Z2))**2

# Calculate derivative of Z2 with respect to A1
_Z2 = A1*W2 + B2
dZ2_dA1 = diff(_Z2, A1)
print("dZ2_dA1 = ", dZ2_dA1) # W2

# Calculate derivative of Z2 with respect to W2
dZ2_dW2 = diff(_Z2, W2)
print("dZ2_dW2 = ", dZ2_dW2) # A1

# Calculate derivative of Z2 with respect to B2
dZ2_dB2 = diff(_Z2, B2)
print("dZ2_dB2 = ", dZ2_dB2) # 1

# Calculate derivative of A1 with respect to Z1
relu = lambda x: Max(x, 0)
_A1 = relu(Z1)
```

```
d_relu = lambda x: x > 0 # Slope is 1 if positive, 0 if negative
dA1_dZ1 = d_relu(Z1)
print("dA1_dZ1 = ", dA1_dZ1) # Z1 > 0

# Calculate derivative of Z1 with respect to W1
_Z1 = X*W1 + B1
dZ1_dW1 = diff(_Z1, W1)
print("dZ1_dW1 = ", dZ1_dW1) # X

# Calculate derivative of Z1 with respect to B1
dZ1_dB1 = diff(_Z1, B1)
print("dZ1_dB1 = ", dZ1_dB1) # 1
```

Notice that ReLU was calculated manually rather than using SymPy's diff() function. This is because derivatives work with smooth curves, not jagged corners that exist on ReLU. But it's easy to hack around that simply by declaring the slope to be 1 for positive numbers and 0 for negative numbers. This makes sense because negative numbers have a flat line with slope 0. But positive numbers are left as is with a 1-to-1 slope.

These partial derivatives can be chained together to create new partial derivatives with respect to the weights and biases. Let's get all four partial derivatives for the weights in W_1, W_2, B_1, and B_2 with respect to the cost function. We already walked through $\frac{dC}{dW_2}$. Let's show it alongside the other three chained derivatives we need:

$$\frac{dC}{dW_2} = \frac{dZ_2}{dW_2}\frac{dA_2}{dZ_2}\frac{dC}{dA_2} = (A_1)\left(\frac{e^{-Z_2}}{\left(1 + e^{-Z_2}\right)^2}\right)(2A_2 - 2y)$$

$$\frac{dC}{dB_2} = \frac{dZ_2}{dB_2}\frac{dA_2}{dZ_2}\frac{dC}{dA_2} = (1)\left(\frac{e^{-Z_2}}{\left(1 + e^{-Z_2}\right)^2}\right)(2A_2 - 2y)$$

$$\frac{dC}{dW_1} = \frac{dC}{DA_2}\frac{DA_2}{dZ_2}\frac{dZ_2}{dA_1}\frac{dA_1}{dZ_1}\frac{dZ_1}{dW_1} = (2A_2 - 2y)\left(\frac{e^{-Z_2}}{\left(1 + e^{-Z_2}\right)^2}\right)(W_2)(Z_1 > 0)(X)$$

$$\frac{dC}{dB_1} = \frac{dC}{DA_2}\frac{DA_2}{dZ_2}\frac{dZ_2}{dA_1}\frac{dA_1}{dZ_1}\frac{dZ_1}{dB_1} = (2A_2 - 2y)\left(\frac{e^{-Z_2}}{\left(1 + e^{-Z_2}\right)^2}\right)(W_2)(Z_1 > 0)(1)$$

We will use these chained gradients to calculate the slope for the cost function C with respect to W_1, B_1, W_2, and B_2.

Automatic Differentiation

As you can see, untangling derivatives even with the chain rule and symbolic libraries like SymPy is still tedious. This is why differentiable programming libraries are rising, such as the JAX library (*https://oreil.ly/N96Pk*) made by Google. It is nearly identical to NumPy except it allows calculating derivatives on parameters packaged as matrices.

If you want to learn more about automatic differentiation, this YouTube video (*https://youtu.be/wG_nF1awSSY*) does a great job explaining it.

Stochastic Gradient Descent

We are now ready to integrate the chain rule to perform stochastic gradient descent. To keep things simple, we are going to sample only one training record on every iteration. Batch and mini-batch gradient descent are commonly used in neural networks and deep learning, but there's enough linear algebra and calculus to juggle just one sample per iteration.

Let's take a look at our full implementation of our neural network, with backpropagated stochastic gradient descent, in Example 7-11.

Example 7-11. Implementing a neural network using stochastic gradient descent

```
import numpy as np
import pandas as pd
from sklearn.model_selection import train_test_split

all_data = pd.read_csv("https://tinyurl.com/y2qmhfsr")

# Learning rate controls how slowly we approach a solution
# Make it too small, it will take too long to run.
# Make it too big, it will likely overshoot and miss the solution.
L = 0.05

# Extract the input columns, scale down by 255
all_inputs = (all_data.iloc[:, 0:3].values / 255.0)
all_outputs = all_data.iloc[:, -1].values

# Split train and test data sets
X_train, X_test, Y_train, Y_test = train_test_split(all_inputs, all_outputs,
    test_size=1 / 3)
n = X_train.shape[0]
```

```python
# Build neural network with weights and biases
# with random initialization
w_hidden = np.random.rand(3, 3)
w_output = np.random.rand(1, 3)

b_hidden = np.random.rand(3, 1)
b_output = np.random.rand(1, 1)

# Activation functions
relu = lambda x: np.maximum(x, 0)
logistic = lambda x: 1 / (1 + np.exp(-x))

# Runs inputs through the neural network to get predicted outputs
def forward_prop(X):
    Z1 = w_hidden @ X + b_hidden
    A1 = relu(Z1)
    Z2 = w_output @ A1 + b_output
    A2 = logistic(Z2)
    return Z1, A1, Z2, A2

# Derivatives of Activation functions
d_relu = lambda x: x > 0
d_logistic = lambda x: np.exp(-x) / (1 + np.exp(-x)) ** 2

# returns slopes for weights and biases
# using chain rule
def backward_prop(Z1, A1, Z2, A2, X, Y):
    dC_dA2 = 2 * A2 - 2 * Y
    dA2_dZ2 = d_logistic(Z2)
    dZ2_dA1 = w_output
    dZ2_dW2 = A1
    dZ2_dB2 = 1
    dA1_dZ1 = d_relu(Z1)
    dZ1_dW1 = X
    dZ1_dB1 = 1

    dC_dW2 = dC_dA2 @ dA2_dZ2 @ dZ2_dW2.T

    dC_dB2 = dC_dA2 @ dA2_dZ2 * dZ2_dB2

    dC_dA1 = dC_dA2 @ dA2_dZ2 @ dZ2_dA1

    dC_dW1 = dC_dA1 @ dA1_dZ1 @ dZ1_dW1.T

    dC_dB1 = dC_dA1 @ dA1_dZ1 * dZ1_dB1

    return dC_dW1, dC_dB1, dC_dW2, dC_dB2

# Execute gradient descent
for i in range(100_000):
    # randomly select one of the training data
```

```
    idx = np.random.choice(n, 1, replace=False)
    X_sample = X_train[idx].transpose()
    Y_sample = Y_train[idx]

    # run randomly selected training data through neural network
    Z1, A1, Z2, A2 = forward_prop(X_sample)

    # distribute error through backpropagation
    # and return slopes for weights and biases
    dW1, dB1, dW2, dB2 = backward_prop(Z1, A1, Z2, A2, X_sample, Y_sample)

    # update weights and biases
    w_hidden -= L * dW1
    b_hidden -= L * dB1
    w_output -= L * dW2
    b_output -= L * dB2

# Calculate accuracy
test_predictions = forward_prop(X_test.transpose())[3]  # grab only A2
test_comparisons = np.equal((test_predictions >= .5).flatten().astype(int), Y_test)
accuracy = sum(test_comparisons.astype(int) / X_test.shape[0])
print("ACCURACY: ", accuracy)
```

There is a lot going on here, but it builds on everything else we learned in this chapter. We perform 100,000 iterations of stochastic gradient descent. Splitting the training and testing data by 2/3 and 1/3, respectively, I get approximately 97–99% accuracy in my testing dataset depending on how the randomness works out. This means after training, my neural network correctly identifies 97–99% of the testing data with the right light/dark font predictions.

The backward_prop() function is key here, implementing the chain rule to take the error in the output node (the squared residual), and then divide it up and distribute it backward to the output and hidden weights/biases to get the slopes with respect to each weight/bias. We then take those slopes and nudge the weights/biases in the for loop, respectively, multiplying with the learning rate L just like we did in Chapters 5 and 6. We do some matrix-vector multiplication to distribute the error backward based on the slopes, and we transpose matrices and vectors when needed so the dimensions between rows and columns match up.

If you want to make the neural network a bit more interactive, here's a snippet of code in Example 7-12 where we can type in different background colors (through an R, G, and B value) and see if it predicts a light or dark font. Append it to the bottom of the previous code Example 7-11 and give it a try!

Example 7-12. Adding an interactive shell to our neural network

```
# Interact and test with new colors
def predict_probability(r, g, b):
    X = np.array([[r, g, b]]).transpose() / 255
    Z1, A1, Z2, A2 = forward_prop(X)
    return A2

def predict_font_shade(r, g, b):
    output_values = predict_probability(r, g, b)
    if output_values > .5:
        return "DARK"
    else:
        return "LIGHT"

while True:
    col_input = input("Predict light or dark font. Input values R,G,B: ")
    (r, g, b) = col_input.split(",")
    print(predict_font_shade(int(r), int(g), int(b)))
```

Building your own neural network from scratch is a lot of work and math, but it gives you insight into their true nature. By working through the layers, the calculus, and the linear algebra, we get a stronger sense of what deep learning libraries like PyTorch and TensorFlow do behind the scenes.

As you have gathered from reading this entire chapter, there are a lot of moving parts to make a neural network tick. It can be helpful to put a breakpoint in different parts of the code to see what each matrix operation is doing. You can also port the code into a Jupyter Notebook to get more visual insight into each step.

3Blue1Brown on Backpropagation

3Blue1Brown has some classic videos talking about backpropagation (*https://youtu.be/Ilg3gGewQ5U*) and the calculus behind neural networks (*https://youtu.be/tIeHLnjs5U8*).

Using scikit-learn

There is some limited neural network functionality in scikit-learn. If you are serious about deep learning, you will probably want to study PyTorch or TensorFlow and get a computer with a strong GPU (there's a great excuse to get that gaming computer you always wanted!). I have been told that all the cool kids are using PyTorch now. However, scikit-learn does have some convenient models available, including the `MLPClassifier`, which stands for "multi-layer perceptron classifier." This is a neural network designed for classification, and it uses a logistic output activation by default.

Example 7-13 is a scikit-learn version of the background color classification application we developed. The `activation` argument specifies which activation function to apply to the nodes contained in the hidden layers.

Example 7-13. Using scikit-learn neural network classifier

```python
import pandas as pd
# load data
from sklearn.model_selection import train_test_split
from sklearn.neural_network import MLPClassifier

df = pd.read_csv('https://bit.ly/3GsNzGt', delimiter=",")

# Extract input variables (all rows, all columns but last column)
# Note we should do some linear scaling here
X = (df.values[:, :-1] / 255.0)

# Extract output column (all rows, last column)
Y = df.values[:, -1]

# Separate training and testing data
X_train, X_test, Y_train, Y_test = train_test_split(X, Y, test_size=1/3)

nn = MLPClassifier(solver='sgd',
                   hidden_layer_sizes=(3, ),
                   activation='relu',
                   max_iter=100_000,
                   learning_rate_init=.05)

nn.fit(X_train, Y_train)

# Print weights and biases
print(nn.coefs_ )
print(nn.intercepts_)

print("Training set score: %f" % nn.score(X_train, Y_train))
print("Test set score: %f" % nn.score(X_test, Y_test))
```

Running this code I get about 99.3% accuracy on my testing data.

MNIST Example Using scikit-learn

To see a scikit-learn example predicting handwritten digits using the MNIST dataset, turn to Appendix A.

Limitations of Neural Networks and Deep Learning

For all of their strengths, neural networks struggle with certain types of tasks. This flexibility with layers, nodes, and activation functions makes it flexible fitting to data in a nonlinear manner...probably too flexible. Why? It can overfit to the data. Andrew Ng, a pioneer in deep learning education and the former head of Google Brain, mentioned this as a problem in a press conference in 2021. Asked why machine learning has not replaced radiologists yet, this was his answer in an *IEEE Spectrum* article (*https://oreil.ly/ljXsz*):

> It turns out that when we collect data from Stanford Hospital, then we train and test on data from the same hospital, indeed, we can publish papers showing [the algorithms] are comparable to human radiologists in spotting certain conditions.

> It turns out [that when] you take that same model, that same AI system, to an older hospital down the street, with an older machine, and the technician uses a slightly different imaging protocol, that data drifts to cause the performance of AI system to degrade significantly. In contrast, any human radiologist can walk down the street to the older hospital and do just fine.

> So even though at a moment in time, on a specific data set, we can show this works, the clinical reality is that these models still need a lot of work to reach production.

In other words, the machine learning overfitted to the Stanford hospital training and testing dataset. When taken to other hospitals with different machinery, the performance degraded significantly due to the overfitting.

The same challenges occur with autonomous vehicles and self-driving cars. It is not enough to just train a neural network on one stop sign! It has to be trained on countless combinations of conditions around that stop sign: good weather, rainy weather, night and day, with graffiti, blocked by a tree, in different locales, and so on. In traffic scenarios, think of all the different types of vehicles, pedestrians, pedestrians dressed in costumes, and infinite number of edge cases that will be encountered! There is simply no effective way to capture every type of event that is encountered on the road just by having more weights and biases in a neural network.

This is why autonomous vehicles themselves do not use neural networks in an end-to-end manner. Instead, different software and sensor modules are broken up where one module may use a neural network to draw a box around an object. Then another module will use a different neural network to classify the object in that box, such as a pedestrian. From there, traditional rule-based logic will attempt to predict the path of the pedestrian and hardcoded logic will choose from different conditions on how to react. The machine learning was limited to label-making activity, not the tactics and maneuvers of the vehicle. On top of that, basic sensors like radar will simply stop if an unknown object is detected in front of the vehicle, and this is just another piece of the technology stack that does not use machine learning or deep learning.

This might be surprising given all the media headlines about neural networks and deep learning beating humans in games like Chess and Go (*https://oreil.ly/9zFxM*), or even besting pilots in combat flight simulations (*https://oreil.ly/hbdYI*). It is important to remember in reinforcement learning environments like these that simulations are closed worlds, where infinite amounts of labeled data can be generated and learned through a virtual finite world. However, the real world is not a simulation where we can generate unlimited amounts of data. Also, this is not a philosophy book so we will pass on discussions whether we live in a simulation. Sorry, Elon! Collecting data in the real world is expensive and hard. On top of that, the real world is filled with infinite unpredictability and rare events. All these factors drive machine learning practitioners to resort to data entry labor to label pictures of traffic objects (*https://oreil.ly/mhjvz*) and other data. Autonomous vehicle startups often have to pair this kind of data entry work with simulated data, because the miles and edge case scenarios needed to generate training data are too astronomical to gather simply by driving a fleet of vehicles millions of miles.

These are all reasons why AI research likes to use board games and video games, because unlimited labeled data can be generated easily and cleanly. Francis Chollet, a renowned engineer at Google who developed Keras for TensorFlow (and also wrote a great book, *Deep Learning with Python*), shared some insight on this in a *Verge* article (*https://oreil.ly/4PDLf*):

> The thing is, once you pick a measure, you're going to take whatever shortcut is available to game it. For instance, if you set chess-playing as your measure of intelligence (which we started doing in the 1970s until the 1990s), you're going to end up with a system that plays chess, and that's it. There's no reason to assume it will be good for anything else at all. You end up with tree search and minimax, and that doesn't teach you anything about human intelligence. Today, pursuing skill at video games like Dota or StarCraft as a proxy for general intelligence falls into the exact same intellectual trap…
>
> If I set out to "solve" Warcraft III at a superhuman level using deep learning, you can be quite sure that I will get there as long as I have access to sufficient engineering talent and computing power (which is on the order of tens of millions of dollars for a task like this). But once I'd have done it, what would I have learned about intelligence or generalization? Well, nothing. At best, I'd have developed engineering knowledge about scaling up deep learning. So I don't really see it as scientific research because it doesn't teach us anything we didn't already know. It doesn't answer any open question. If the question was, "Can we play X at a superhuman level?," the answer is definitely, "Yes, as long as you can generate a sufficiently dense sample of training situations and feed them into a sufficiently expressive deep learning model." We've known this for some time.

That is, we have to be careful to not conflate an algorithm's performance in a game with broader capabilities that have yet to be solved. Machine learning, neural networks, and deep learning all work narrowly on defined problems. They cannot broadly reason or choose their own tasks, or ponder objects they have not seen before. Like any coded application, they do only what they were programmed to do.

With whatever tool it takes, solve the problem. There should be no partiality to neural networks or any other tool at your disposal. With all of this in mind, using a neural network might not be the best option for the task in front of you. It is important to always consider what problem you are striving to solve without making a specific tool your primary objective. The use of deep learning has to be strategic and warranted. There are certainly use cases, but in most of your everyday work you will likely have more success with simpler and more biased models like linear regression, logistic regression, or traditional rule-based systems. But if you find yourself having to classify objects in images, and you have the budget and labor to build that dataset, then deep learning is going to be your best bet.

Is an AI Winter Coming?

Are neural networks and deep learning useful? Absolutely! And they are definitely worth learning. That said, there is a good chance you have seen media, politicians, and tech celebrities hail deep learning as some general artificial intelligence that can match, if not outperform, human intelligence and is even destined to control the world. I have attended talks from authoritative figures in the software development community telling programmers they will be out of a job in a few years because of machine learning, and AI will take over writing code.

Nearly a decade later, these predictions are still running late in 2022 because they are simply not true. We have seen far more AI challenges than breakthroughs, and I still have to drive my own car and write my own code. Neural networks are *loosely* inspired by the human brain but are by no means a replication of them. Their capabilities are nowhere near on par with what you see in movies like *The Terminator*, *Westworld*, or *WarGames*. Instead, neural networks and deep learning work narrowly on specific problems, like recognizing dog and cat photos after being optimized on thousands of images. As stated earlier, they cannot reason or choose their own tasks, or contemplate uncertainty or objects they have not seen before. Neural networks and deep learning do only what they were programmed to do.

This disconnect may have inflated investment and expectations, resulting in a bubble that could burst. This would bring about another "AI winter," where disillusionment and disappointment dry up funding in AI research. In North America, Europe, and Japan, AI winters have happened multiple times since the 1960s. There's a good chance another AI winter is around the corner, but it does not mean neural networks and deep learning will lose usefulness. They will continue to be applied in the problems they are good at: computer vision, audio, natural language, and a few other domains. Maybe you can discover new ways to use them! Use what works best whether it's linear regression, logistic regression, a traditional rule-based system, or a neural network. There is much more power in the simplicity of pairing the right tool to the right problem.

Conclusion

Neural networks and deep learning offer some exciting applications, and we just scratched the surface in this one chapter. From recognizing images to processing natural language, there continue to be use cases for applying neural networks and their different flavors of deep learning.

From scratch, we learned how to structure a simple neural network with one hidden layer to predict whether or not a light or dark font should be used against a background color. We also applied some advanced calculus concepts to calculate partial derivatives of nested functions and applied it to stochastic gradient descent to train our neural network. We also touched on libraries like scikit-learn. While we do not have the bandwidth in this book to talk about TensorFlow, PyTorch, and more advanced applications, there are great resources out there to expand your knowledge.

3Blue1Brown has a fantastic playlist on neural networks and backpropagation (*https://oreil.ly/VjwBr*), and it is worth watching multiple times. Josh Starmer's StatQuest playlist on neural networks (*https://oreil.ly/YWnF2*) is helpful as well, particularly in visualizing neural networks as manifold manipulation. Another great video about manifold theory and neural networks can be found here on the Art of the Problem (*https://youtu.be/e5xKayCBOeU*). Finally, when you are ready to deep-dive, check out Aurélien Géron's *Hands-On Machine Learning with Scikit-Learn, Keras, and TensorFlow* (O'Reilly) and Francois Chollet's *Deep Learning with Python* (Manning).

If you made it to the end of this chapter and feel you absorbed everything within reason, congrats! You have not just effectively learned probability, statistics, calculus, and linear algebra but also applied it to practical applications like linear regression, logistic regression, and neural networks. We will talk about how you can proceed in the next chapter and start a new phase of your professional growth.

Exercise

Apply a neural network to the employee retention data we worked with in Chapter 6. You can import the data here (*https://tinyurl.com/y6r7qjrp*). Try to build this neural network so it predicts on this dataset and use accuracy and confusion matrices to evaluate performance. Is it a good model for this problem? Why or why not?

While you are welcome to build the neural network from scratch, consider using scikit-learn, PyTorch, or another deep learning library to save time.

Answers are in Appendix B.

## CHAPTER 8

Career Advice and the Path Forward

As we come to a close with this book, it is a good idea to evaluate where to go from here. You learned and integrated a wide survey of applied mathematical topics: calculus, probability, statistics, and linear algebra. Then you applied these techniques to practical applications, including linear regression, logistic regression, and neural networks. In this chapter, we will cover how to use this knowledge going forward while navigating the strange, exciting, and oddly diverse landscape of a data science career. I will emphasize the importance of having direction and a tangible objective to work toward, rather than memorizing tools and techniques without an actual problem in mind.

Since we're moving away from foundational concepts and applied methods, this chapter will have a different tone than the rest of the book. You might be expecting to learn how you can apply these mathematical modeling skills to your career in focused and tangible ways. However, if you want to be successful in a data science career, you will have to learn a few more hard skills like SQL and programming, as well as soft skills to develop professional awareness. The latter are especially important so you do not become lost in the shape-shifting profession that is data science and unseen market forces blindside you.

I am not going to presume to know your career goals or what you hope to achieve with this information. I will make a few safe bets, though, since you are reading this book. I imagine you might be interested in a data science career, or you have worked in data analysis and desire to formalize your analytics knowledge. Perhaps you come from a software engineering background, seeking to get a grasp of AI and machine learning. Maybe you are a project manager of some kind and find you need to understand the capabilities of a data science or AI team so you can scope accordingly. Maybe you are just a curious professional wondering how math can be useful on a practical level, not just an academic one.

I will do my best to meet the concerns of all these groups and hopefully generalize some career advice that will be useful to most readers. Let's start with redefining data science. We have objectively studied it and will now look at it in the context of career development and the future of the field.

Is the Author Being Anecdotal?

It is hard to not sound anecdotal in tasks like this, where I dispense career advice sharing my own experiences (and experiences of others) rather than large, controlled surveys and studies like I advocate in Chapter 3. But here's my pitch: I have worked in the Fortune 500 world for more than a decade and saw the transformation that the data science movement had on organizations. I have spoken at many tech conferences around the world and listened to countless peers react with "this has happened to us too!" I read many blogs and respected publications, from the *Wall Street Journal* to *Forbes*, and I have learned to recognize disconnects between popular expectations and reality. I pay especially close attention to the kingmakers, leaders, and followers across different industries and how they make or follow markets with data science and artificial intelligence. Currently, I teach and advise stakeholders in safety-critical applications of artificial intelligence at University of Southern California, in the Aviation Safety and Security Program.

I only put a CV here to show that while I have not done formal surveys and studies, and perhaps I am compiling anecdotal information, there are persistent narratives I have found with all these sources of information. Vicki Boykis, an astute machine learning engineer at Tumblr, wrote a blog article sharing similar findings to mine (*https://oreil.ly/vm8Vp*), and I highly recommend you read it. You are welcome to take my findings with a grain of salt, but pay close attention to what is happening in your own work environment and identify assumptions that your management and colleagues subscribe to.

Redefining Data Science

Data science is the analysis of data to gain actionable insights. In practice, it is an amalgamation of different data-related disciplines rolled into one: statistics, data analysis, data visualization, machine learning, operations research, software engineering... just to name a few. Pretty much any discipline that touches data can be branded as "data science." This lack of clear definition has been problematic for the field. After all, anything that lacks definition is wide open to interpretability, like a piece of abstract art. This is why HR departments struggle with "data scientist" job postings, as they tend to be all over the map (*https://oreil.ly/NHnbu*). Figure 8-1 shows an umbrella encompassing different disciplines and tools that can fall under data science.

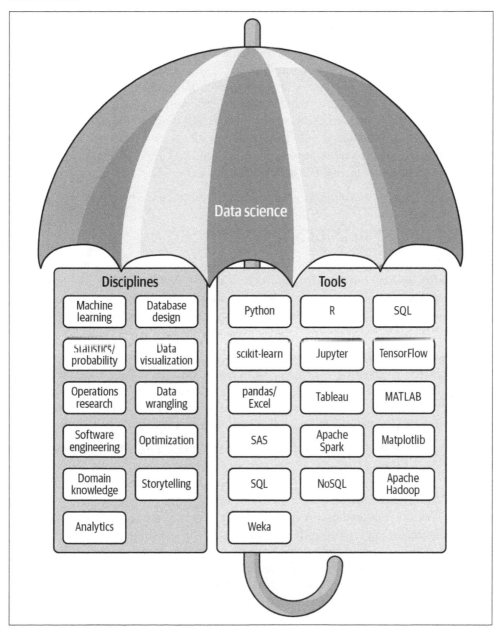

Figure 8-1. The data science umbrella

How did we get here? And how can something lacking definition like data science become such a compelling force in the corporate world? Most important, how does its definition (or lack thereof) affect your career? These are all important questions we talk about in this chapter.

This is why I tell my clients a better definition for *data science* is software engineering with proficiency in statistics, machine learning, and optimization. If you take any of those four disciplines away (software engineering, statistics, machine learning, and optimization), the data scientist is at risk of not performing. Most organizations have struggled to make clear distinctions about skillsets that make a data scientist effective, but the definition provided should give clarity. While some may think that software engineering is a controversial requirement, I think it is highly necessary given the direction industries are going. We will address this point later.

But first, the best way to understand data science is to trace the history of the term.

A Brief History of Data Science

We could trace data science all the way back to the origin of statistics, as early as the 17th century or even the 8th century (*https://oreil.ly/tYPB5*). For the sake of brevity, let's start in the 1990s. Analysts, statisticians, researchers, "quants," and data engineers often maintained distinct roles. Tooling stacks often consisted of spreadsheets, R, MATLAB, SAS, and SQL.

Of course, things started to change rapidly after 2000. The internet and connected devices started to generate enormous amounts of data. Alongside the inception of Hadoop, Google pushed analytics and data collection to unimaginable heights. As 2010 approached, executives at Google insisted statisticians will have the "sexy" job in the next decade (*https://oreil.ly/AZgfM*). What came next was prescient.

In 2012, the *Harvard Business Review* mainstreamed this concept called data science and declared it the "sexiest job of the 21st century" (*https://oreil.ly/XYbrf*). After the *Harvard Business Review* article, many companies and corporate workers raced to fill the data science void. Management consultants were primed and positioned to educate Fortune 500 leaders on how to bring data science to their organizations. SQL developers, analysts, researchers, quants, statisticians, engineers, physicists, and a myriad of other professionals rebranded themselves as "data scientists." Tech companies, feeling that traditional role titles like "analyst," "statistician," or "researcher" sounded dated, renamed the roles to "data scientist."

Naturally, management at Fortune 500 companies came under pressure from the C-suite to get on the data science bandwagon. The initial reasoning was a lot of data was being collected, therefore big data was becoming a trend and data scientists were needed to get insight from it. Around this time, the word "data-driven" became a maxim across industries. The corporate world believed that unlike people, data is objective and unbiased.

Reminder: Data Is Not Objective and Unbiased!

To this day, many professionals and managers fall into the fallacy of thinking data is objective and unbiased. Hopefully, after reading this book you know this is simply not true, and refer to Chapter 3 if you need a refresher about why.

Management and HR at companies, unable to compete for specialized deep learning PhDs that were gobbled up by FAANG (Facebook, Amazon, Apple, Netflix, and Google) and still under pressure to check the data science box, made an interesting move. They rebranded existing teams of analysts, SQL developers, and Excel jockeys as "data scientists." Cassie Kozyrkov, the Chief Decision Scientist at Google, described this open secret in a 2018 blog article on Hackernoon (*https://oreil.ly/qNl53*):

> With every *data scientist* title I've held, I had already been doing the job under a different name before rebranding czars in HR applied a little nip-tuck to the employee database. My duties didn't change in the slightest. I'm no exception; my social circle is full of former statisticians, decision support engineers, quantitative analysts, math professors, big data specialists, business intelligence experts, analytics leads, research scientists, software engineers, Excel jockeys, niche PhD survivors...all proud Data Scientists of today.

Data science technically does not exclude any of these professionals, because they all are using data to gather insight. Of course, there was some pushback from the science community, which was reluctant to coronate data science as a real science. After all, can you think of a science that does not use data? In 2011 Pete Warden (who now works as a TensorFlow lead at Google) wrote an interesting defense of the data science movement in an O'Reilly article (*https://oreil.ly/HXgvI*). He also cleanly articulated objectors' arguments about the lack of definition:

> [Regarding the lack of definition for data science, this] is probably the deepest objection, and the one with the most teeth. There is no widely accepted boundary for what's inside and outside of data science's scope. Is it just a faddish rebranding of statistics? I don't think so, but I also don't have a full definition. I believe that the recent abundance of data has sparked something new in the world, and when I look around I see people with shared characteristics who don't fit into traditional categories. These people tend to work beyond the narrow specialties that dominate the corporate and institutional world, handling everything from finding the data, processing it at scale, visualizing it and writing it up as a story. They also seem to start by looking at what the data can tell them, and then picking interesting threads to follow, rather than the traditional scientist's approach of choosing the problem first and then finding data to shed light on it.

Pete ironically could not come up with a definition for data science either, but he clearly made a case for why data science is flawed but useful. He also highlighted the shift of research abandoning the scientific method in favor of once-eschewed practices like data mining, which we talked about in Chapter 3.

Data science made an interesting pivot just a few years after the *Harvard Business Review* article. Perhaps it was more of a merger with AI and machine learning than a pivot. Regardless, when machine learning and deep learning dominated headlines around 2014, data was sold as the "fuel" for creating artificial intelligence. This naturally expanded the scope of data science and created a confluence with the AI/machine learning movement. In particular, the ImageNet challenge sparked a revived interest in AI and brought on a machine learning and deep learning renaissance. Companies like Waymo and Tesla promised self-driving cars in a matter of years thanks to advances in deep learning, further fueling media headlines and bootcamp enrollments.

This sudden interest in neural networks and deep learning had a interesting side effect. Regression techniques like decision trees, support vector machines, and logistic regression, which has spent decades hiding in academia and specialized statistics professions, rode the coattails of deep learning into the public limelight. At the same time, libraries like scikit-learn created a low barrier for entry into the field. This had a hidden cost of creating data science professionals who did not understand how these libraries or models worked but used them anyway.

Because the discipline of data science has evolved faster than a perceived need to define it, it is not uncommon for a data scientist role to be a complete wildcard. I have met several individuals who held a data scientist title in Fortune 500 companies. Some are highly proficient in coding and might even have software engineering backgrounds but have no idea what statistical significance is. Others stay confined in Excel and barely know SQL, much less Python or R. I met data science folks who taught themselves a few functions in scikit-learn and quickly found themselves floundering because that was all they knew.

So what does this mean for you? How do you flourish in such a buzzwordy and chaotic landscape? It all comes down to what types of problems or industries interest you, and not being quick to rely on employers to define roles. You do not have to be a data scientist to do data science. There are a wide array of fields you can work to your advantage given this knowledge you now have. You can be an analyst, researcher, machine learning engineer, advisor, consultant, and a myriad of other roles that are not necessarily called data scientist.

But first, let's address some ways you can continue learning and find your edge in the data science job market.

Finding Your Edge

The practical data science professional needs more than an understanding of statistics and machine learning to thrive. In most cases, it is unreasonable to expect data to be readily available for machine learning and other projects. Instead, you will find yourself chasing data sources, engineering scripts and software, scraping documents, scraping Excel workbooks, and even creating your own databases. At least 95% of your coding efforts will not be related to machine learning or statistical modeling at all, but rather creating, moving, and transforming data so it is usable.

On top of that, you have to be aware of the big picture and the dynamics of your organization. Your managers can make assumptions in defining your role, and it is important to identify these assumptions so you recognize how they affect you. While you are relying on your customers and leadership for their industry expertise, you should be in a role to provide technical knowledge and articulate what is feasible. Let's look at some hard and soft skills you will likely need.

SQL Proficiency

SQL, also called *structured query language*, is a querying language to retrieve, transform, and write table data. A *relational database* is the most common way to organize data, storing data into tables that are connected to each other much like a VLOOKUP in Excel. Relational database platforms like MySQL, Microsoft SQL Server, Oracle, SQLite, and PostgreSQL support SQL. As you might notice, SQL and relational databases are so tightly coupled that "SQL" is often used in the branding of the relational database, like "MySQL" and "Microsoft SQL Server."

Example 8-1 is a simple SQL query that retrieves the CUSTOMER_ID and NAME fields from a CUSTOMER table, for records that are in the STATE of 'TX'.

Example 8-1. A simple SQL query

```
SELECT CUSTOMER_ID, NAME
FROM CUSTOMER
WHERE STATE = 'TX'
```

Simply put, it is hard to get anywhere as a data science professional without proficiency in SQL. Businesses use data warehouses and SQL is almost always the means to retrieve the data. SELECT, WHERE, GROUP BY, ORDER BY, CASE, INNER JOIN, and LEFT JOIN should all be familiar SQL keywords. It is even better to know subqueries, derived tables, common table expressions, and windowing functions to get the most utility out of your data.

Shameless Plug: The Author Has an SQL Book!

I have written a beginner's SQL book for O'Reilly called *Getting Started with SQL* (*https://oreil.ly/K2Na9*). It is barely over one hundred pages so it can be finished in a day. It covers the essentials, including joins and aggregations as well as creating your own database. The book uses SQLite, which can be set up in less than a minute.

There are other great SQL books by O'Reilly, including *Learning SQL, 3rd Ed.* by Alan Beaulieu and the *SQL Pocket Guide, 4th Ed.* by Alice Zhao. After speed-reading my one-hundred-page primer, check out both of these books too.

SQL is also critical in making Python or other programming languages easily talk to databases. If you want to send SQL queries to a database from Python, you can bring the data back as Pandas DataFrames, Python collections, and other structures.

Example 8-2 shows a simple SQL query run in Python using the SQLAlchemy library. It returns the records as named tuples. Just be sure to download this SQLite database file (*https://bit.ly/3F8heTS*) and place it in your Python project, as well as run `pip install sqlalchemy`.

Example 8-2. Running an SQL query within Python using SQLAlchemy

```
from sqlalchemy import create_engine, text

engine = create_engine('sqlite:///thunderbird_manufacturing.db')
conn = engine.connect()

stmt = text("SELECT * FROM CUSTOMER")
results = conn.execute(stmt)

for customer in results:
    print(customer)
```

What About Pandas and NoSQL?

I frequently get questions regarding "alternatives" to SQL such as NoSQL or Pandas. These really are not alternatives and instead are different tools residing elsewhere in the data science toolchain. Take Pandas, for example. In Example 8-3, I can create an SQL query pulling all records from the table CUSTOMER and put them in a Pandas DataFrame.

Example 8-3. Importing an SQL query into a Pandas DataFrame

```
from sqlalchemy import create_engine, text
import pandas as pd

engine = create_engine('sqlite:///thunderbird_manufacturing.db')
conn = engine.connect()

df = pd.read_sql("SELECT * FROM CUSTOMER", conn)
print(df) # prints SQL results as DataFrame
```

SQL was used here to bridge the gap between the relational database and my Python environment, and load the data into a Pandas `DataFrame`. If I have demanding computations that SQL is equipped to handle, it is more efficient that I do that on the database server using SQL rather than locally on my computer using Pandas. Simply put, Pandas and SQL can work together and are not competing technologies.

It is the same thing with NoSQL, which includes platforms like Couchbase and MongoDB. While some readers might disagree and make valid arguments, I believe comparing NoSQL with SQL is comparing apples and oranges. Yes, they both store data and provide querying capabilities, but I do not think that puts them in competition. They have different qualities for different use cases. NoSQL stands for "not only SQL" and is better equipped to store unstructured data, like pictures or free-form text articles. SQL is better equipped to store structured data. SQL maintains data integrity more aggressively than NoSQL, although at the cost of computing overhead and less scalability.

SQL, the Lingua Franca of Data

In 2015, many speculated NoSQL and distributed data processing technologies like Apache Spark would replace SQL and relational databases. Ironically, SQL has proven to be so important to data users that SQL layers were added to these platforms due to demand. These layering technologies include Presto (*https://oreil.ly/Qf6c1*), BigQuery (*https://oreil.ly/iCEWW*), and Apache Spark SQL (*https://oreil.ly/IIPft*), to name a few. Most data problems are not big data problems, and nothing works quite like SQL in querying data. Therefore, SQL has continued to thrive and maintained its place as the lingua franca of the data world.

On another note, the promotion of NoSQL and big data platforms might be a lesson on the Silver Bullet Syndrome. Hadi Hariri from JetBrains gave a talk on this topic in 2015 (*https://oreil.ly/hPEIF*), and it is well worth the watch.

Programming Proficiency

Typically, many data scientists are not skilled at coding, at least not at the level of a software engineer. However, it is becoming increasingly important that they code. This provides an opportunity to gain an edge. Learn object-oriented programming, functional programming, unit tests, version control (e.g., Git and GitHub), Big-O algorithm analysis, cryptography, and other relevant computer science concepts and language features you come across.

Here is why. Let's say you created a promising regression model, like a logistic regression or neural network, based on some sample data you were given. You ask the in-house programmers in your IT department to "plug it in" to an existing piece of software.

They look at your pitch warily. "We need to rewrite this in Java, not Python," they say grudgingly. "Where are your unit tests?" another asks. "Do you not have any classes or types defined? We have to reengineer this code to be object-oriented." On top of that, they do not understand the math of your model and rightfully worry it will misbehave on data it has not seen before. Since you have not defined unit tests, which is not straightforward to do with machine learning, they are unsure how to verify the quality of your model. They also ask how are two versions of code (Python and Java), going to be managed?

You start to feel out of your element, and say, "I don't understand why the Python script cannot just be plugged in." One of them pauses contemplatively and replies, "We could create a web service with Flask and avoid having to rewrite in Java. However, the other concerns do not go away. We then have to worry about scalability and high traffic hitting the web service. Wait...perhaps we can deploy to the Microsoft Azure cloud as a virtual machine scale set, but we still have to architect the backends. Look, this has to be reengineered no matter how you approach it."

This is precisely why many data scientists have work that never leaves their laptop. As a matter of fact, putting machine learning into production has become so elusive it has become a unicorn and hot topic in recent years. There is an enormous gap between data scientists and software engineers, so naturally there is pressure for data science professionals to now be software engineers.

This may sound overwhelming as data science is already overloaded in scope, with many disciplines and requirements. However, this is not to demonstrate that you need to learn Java. You can be an effective software engineer in Python (or whatever employable language you prefer), but you have to be good at it. Learn object-oriented programming, data structures, functional programming, concurrency, and other design patterns. Two good books to tackle these topics for Python include *Fluent Python, 2nd Ed.* by Luciano Ramalho (O'Reilly) and *Beyond the Basic Stuff with Python* by Al Sweigart (No Starch).

After that, learn to solve practical tasks including database APIs, web services (*https://oreil.ly/gN9e7*), JSON parsing (*https://oreil.ly/N8uef*), regular expressions (*https://oreil.ly/IyD2P*), web scraping (*https://oreil.ly/9oWWb*), security and cryptography (*https://oreil.ly/oxliO*), cloud computing (Amazon Web Services, Microsoft Azure), and whatever else helps you become productive in standing up a system.

As stated earlier, the programming language you master does not have to be Python. It can be another language but it is encouraged that the language is universally used and employable. Languages that have higher employability at the time of writing include Python, R, Java, C#, and C++. Swift and Kotlin have a dominant presence on Apple and Android devices, and both are fantastic, well-supported languages. Although many of these languages are not mainstream for data science purposes, it can be helpful to learn at least one other besides Python to get more exposure.

you can create confusing states and bugs that will at best create blatant errors, and at worst create subtle miscalculations that go unnoticed. Especially if you are a beginner, this is a maddening way to learn Python as these technical traps are not obvious to newcomers. It is also a way to come across and present a finding, only to discover it was the result of a bug and therefore too good to be true.

I am not advocating that you avoid notebooks. By all means, use them if they make you and your workplace happy! However I am advocating to not rely on them. Joel Grus, author of *Data Science from Scratch* (O'Reilly), gave a talk at JupyterCon on this very subject that you can watch here (*https://oreil.ly/V00bQ*).

Anchoring Bias and First Programming Languages

It is common for technical professionals to get partial to and emotionally invested in technologies and platforms, especially programming languages. Please don't do this! This kind of tribalism is not productive and ignores the reality that every programming language caters to different qualities and use cases. Another reality is that some programming languages catch on while others do not, often for reasons that have nothing to do with merits in language design. If a big company is not paying for its support, its chances of survival are slim.

We talked about different types of cognitive biases in Chapter 3. Another one is anchoring bias (*https://oreil.ly/sXNh0*), which states that we can become partial to the first thing we learn, such as a programming language. If you are feeling obligated to learn a new language, be open-minded and give it a chance! No language is perfect, and all that matters is it gets the job done.

However, be wary if the language's support is questionable because it is on life support, is not receiving updates, or lacks a corporate maintainer. Examples of this include Microsoft's VBA (*https://oreil.ly/B8c5A*), Red Hat's Ceylon (*https://oreil.ly/LJdw4*), and Haskell (*https://oreil.ly/ASnnN*).

Java Data Science Libraries

While not nearly as popular as Python data science libraries, Java does have several equivalent libraries that are strongly supported. ND4J (*https://github.com/deeplearning4j/nd4j*) is the NumPy for the Java virtual machine, and SMILE (*https://haifengl.github.io*) is the scikit-learn. TableSaw (*https://github.com/jtablesaw/tablesaw*) is the Java equivalent to Pandas.

Apache Spark (*https://spark.apache.org*) is actually written on the Java platform, specifically using Scala. Interestingly, Apache Spark was a driver for some time trying to make Scala a mainstream language for data science, although it did not catch on to the degree the Scala community probably hoped. This is why a lot of effort was made to add Python, SQL, and R compatibility and not just Java and Scala.

Data Visualization

Another technical skill in which you should have some degree of proficiency is data visualization. Be comfortable making charts, graphs, and plots that not only tell stories to management but also help your own data exploration efforts. You can summarize data with an SQL command, but sometimes a bar chart or scatterplot will give you a better sense of your data in less time.

When it comes to what tools to use for data visualization, this is much harder to answer because there is so much fragmentation and choice. If you work in a traditional office environment, Excel and PowerPoint are often the preferred visualization tools, and you know what? They are quite all right! I do not use them for everything but they do accomplish a great majority of tasks. Need a scatterplot on a small/medium dataset? Or a histogram? No problem! You can have one built in a few minutes after copying/pasting your data into an Excel workbook. This is great for one-time graph visualizations, and there is no shame in using Excel when it works.

However, there are situations where you might want to script the creation of graphs so it is repeatable and reusable or it integrates with your Python code. matplotlib (*https://matplotlib.org*) has been the go-to for some time, and it is hard to avoid when Python is your platform. Seaborn (*https://seaborn.pydata.org*) provides a wrapper on top of matplotlib to make it easier to use for common chart types. SymPy, which we used a lot throughout this book, uses matplotlib as its backend. However, some consider matplotlib to be so mature it is approaching legacy status. Libraries like Plotly (*https://plotly.com/python*) have been on the rise and are nice to use, and it is based on the JavaScript D3.js library (*https://d3js.org*). Personally, I am having success with Manim (*https://www.manim.community*). The 3Blue1Brown-style visualizations it produces are extraordinary and have that "Wow!" factor with customers, and the API is surprisingly easy to use considering the animation power it has. However, it is a young library and has yet to reach maturity, meaning you might have breaking code changes as it evolves with each release.

You cannot go wrong with exploring all of these solutions, and if your employer/customer does not have a preference, you can find one that works best for you.

There are commercial licensed platforms like Tableau (*https://www.tableau.com/prod ucts/desktop*), which are fine to a degree. They set out to create proprietary software that specializes in visualization and create a drag-and-drop interface so it is accessible to nontechnical users. Tableau even has a whitepaper titled "Make Everyone in Your Organization a Data Scientist" (*https://oreil.ly/kncmP*), which does not help the data scientist definition problem mentioned earlier. The challenges I have found with Tableau is it only does visualization well and requires a hefty license. While you can somewhat integrate Python with TabPy (*https://tableau.github.io/TabPy/docs/about.html*), you might just choose to use the capable open source libraries mentioned earlier unless your employer wants to use Tableau.

Software Licenses Can Be Political

Imagine that you created a Python or Java application that solicits a few user inputs, retrieves and wrangles different data sources, runs some highly custom algorithms, and then presents a visualization and a table showing the result. You present it at a meeting after months of hard work, but then one of the managers raises his hand and asks, "Why not just do this in Tableau?"

It's a hard pill to swallow for some managers, knowing that they have spent thousands of dollars for corporate software licenses, and you come in and use a more capable (albeit more complex to use) open source solution that has no licensing costs. You can emphasize that Tableau does not support these algorithms or integrated workflows you had to create. After all, Tableau is just visualization software. It is not a from-scratch coding platform to create a customized, highly tailored solution.

Leadership is often sold the impression that Tableau, Alteryx, or another commercial tool can do everything. After all, they spent a lot of money for it and were probably given a good sales pitch by the vendor. Naturally, they want to justify the costs and have as many people as possible using the license. They probably spent further budget training employees to use the software and want others to be able to maintain your work.

Be sensitive to this. If management asks you to use the tool they paid for, then explore if you can make it work. But if it has limitations or steep usability compromises in your specific tasks, be politely up-front about it.

Knowing Your Industry

Let's compare two industries: movie streaming (e.g., Netflix) and aerospace defense (e.g., Lockheed Martin). Do they have anything in common? Hardly! Both are technology-driven companies, but one is streaming movies for consumers and the other is building airplanes with ordnance.

When I advise on artificial intelligence and system safety, one of the first things I point out is these two industries have very different tolerances for risk. A movie-streaming company may tout they have an AI system that learns what movies to recommend to consumers, but how catastrophic is it when it gives a bad recommendation? Well, at worst you have a mildly irritated consumer who wasted two hours watching a movie they did not enjoy.

But what about the aerospace defense company? If a fighter jet has AI on board that automatically shoots targets, how catastrophic would it be if it were wrong? We are talking about human lives now, not movie recommendations!

The risk tolerance gap between both of these industries is wide. Naturally, the aerospace defense company is going to be far more conservative implementing any experimental system. This means red tape and safety working groups evaluating and stopping any project they deem unacceptably high risk, and justifiably so. What's interesting, though, is the success of AI in Silicon Valley startups, most in low-risk applications like movie recommendations, have created FOMO ("fear of missing out") with the defense industry executives and leadership. This is likely because the disparity in risk tolerance between these two domains is not highlighted enough.

Of course, there is a wide spectrum of risk severity between these two industries, between "irritated user" and "loss of human life." Banks may use AI to determine who will qualify for loans, but that carries risks in discriminating against certain demographics. Criminal justice systems have experimented with AI in parole and surveillance systems, only to run into the same discrimination issues. Social media may use AI (*https://oreil.ly/VoK95*) to determine what user posts are acceptable but anger its users when it suppresses "harmless" content (false positives) as well as lawmakers when "harmful" content is not suppressed (false negatives).

This demonstrates the need to know your industry. If you want to do lots of machine learning, you will probably want to work in low-risk industries where false positives and false negatives do not endanger or upset anyone. But if none of that appeals to you, and you want to work on bolder applications like self-driving cars (*https://oreil.ly/sOYs6*), aviation, and medicine, then expect your machine learning models to get rejected a lot.

In these higher-risk industries, do not be surprised if specific PhDs and other formal credentials are required. Even with a specialized PhD, false positives and false negatives do not magically disappear. If you do not want to pursue such committed specialization, you will likely be better off learning other tools besides machine learning, including software engineering, optimization, statistics, and business rule systems/heuristics.

Productive Learning

In a 2008 stand-up special, comedian Brian Regan contrasted his lack of curiosity to those who read newspapers. Pointing out that a front-page story is never concluded, he stated he has no desire to turn to the specified page to find out how it ends. "And after a nine-year trial, the jury finally came in with a verdict of continued on page 22 on column C...I guess I'll never know," he quips dismissively. He then contrasts with those who flip the pages, exclaiming, "I want to learn! I want to be a learner of things!"

While Brian Regan might have intended to be self-deprecating, maybe he was right about something. Learning a topic for the sake of it of is hardly motivation, and being disinterested is not always a bad thing. If you pick up a calculus textbook and have no purpose for learning it, you will probably end up discouraged and frustrated. You need to have a project or objective in mind, and if you find a topic uninteresting, why bother learning it? Personally, when I allowed myself to lose interest in topics I could not find relevant, it was incredibly liberating. Even more surprising, my productivity skyrocketed.

It does not mean you should be uncurious. However, there is so much information out there, and prioritizing what you learn is an invaluable skill. You can ask questions about why something is useful and, if you cannot get a straight answer, allow yourself to move on! Is everybody talking about natural language processing? It does not mean you have to! Most businesses do not need natural language processing anyway, so it is OK to say it is not worth your effort or time.

Whether you have projects on your job or you create your own for self-study, have something tangible to work toward. Only *you* get to decide what is worth learning, and you can wave off the FOMO in pursuit of things you find interesting and relevant.

Practitioner Versus Advisor

This might be a generalization, but there are two types of knowledge experts: practitioners and advisors. To find your edge, discern which you want to be and tune your professional development accordingly.

In the data science and analytics world, practitioners are the ones writing code, creating models, scouring data, and trying to directly create value. Advisors are like consultants, telling management whether their objectives are sound, helping with strategy, and providing direction. Sometimes a practitioner can work their way up to being an advisor. Sometimes advisors were never practitioners. There are pros and cons to each role.

A practitioner may enjoy coding, doing data analysis, and performing tangible work that directly can create value. A benefit for the practitioner is they actually develop and possess hard skills. However, burying oneself in code, math, and data makes it easy to lose sight of the big picture and lose touch with the rest of the organization and industry. A common complaint I have heard from managers is their data scientist wants to work on problems *they* find interesting but that do not add value to the organization. I have also heard complaints from practitioners who want exposure and upward mobility but feel tucked away and hidden in their organization.

An advisor admittedly has a cushier job in some ways. They dispense advice and information to managers, and help provide strategic direction to a business. They are typically not the ones who write code or scour data, but they help management hire the people who do. The career risk is different, as they do not worry about meeting sprint deadlines, dealing with code bugs, or misbehaving models like practitioners do. But they do have to worry about staying knowledgeable, credible, and relevant.

To be an effective advisor, you have be *really* knowledgeable and know things other people do not. It has to be critical and relevant information that is fine-tuned to your customer's needs. To stay relevant, you have to read, read, and read more every day...seeking and synthesizing information others are overlooking. It is not enough to be familiar with machine learning, statistics, and deep learning. You have to be paying attention to your client's industry as well as other industries, tracking who is succeeding and who is not. You also have to learn to pair the right solution to the right problem, in a business landscape where many are looking for a silver bullet. And to do all of this, you have to be an effective communicator and share information in a way that helps your client, not just demonstrate what you know.

The greatest risk for an advisor is providing information that ends up being wrong. Some consultants are quite effective at redirecting blame to outside factors, like "nobody in the industry saw this coming" or "this is a six-sigma event!" meaning an undesirable event had a one in a half-billion chance of occurring but did anyway. Another risk is not having the hard skills of a practitioner and becoming disconnected from the technical side of the business. This is why it is a good idea to regularly practice coding and modeling at home, or at least make technical books part of your reading.

In the end, a good advisor tries to be a bridge between the customer and their end goal, often filling a massive knowledge gap that exists. It is not billing the maximum number of hours and inventing busywork, but truly identifying what troubles your customer and helping them sleep at night.

Success Is Not Always About Profitability

Discern how your customer is defining "success." Businesses are pursuing AI, machine learning, and data science to be successful, right? But what is success?

Is it profitability? Not always. In our highly speculative economy, success might be another series of venture capital funding, having customer growth or revenue growth, or high valuation even if the company is losing millions or billions of dollars. None of these metrics has anything to do with profitability.

Why is this happening? Tolerance for long-term plays in venture capital have marginalized profitability, believing it will be achieved much farther down the road. However, it is possible this is creating bubbles much like what happened in the dot-com boom of 2000.

In the end, profitability has to be achieved for a company to be long-term successful, but not everyone has that goal. Many founders and investors simply want to ride the growth and cash out before the bubble bursts, often when the company is sold to the public through an IPO.

What does this mean for you? Whether you are a practitioner or advisor, working at a startup or a Fortune 500, recognize what is motivating your client or employer. Are they seeking higher valuations? Actual profitability? Intrinsic value or perceived value? This will directly affect what you work on and what your clients want to hear, and you can judge accordingly if you are in a place to assist them or not.

If you want to learn more about venture capital, speculative valuation, and the effects of startup culture, *The Cult of We* (*https://www.cultofwe.com*) by *Wall Street Journal* writers Eliot Brown and Maureen Farrell is fantastic.

When projects are planned on tools rather than problems, there is a high likelihood of the project not succeeding. This means as an advisor you have to hone your listening skills and identify the questions that clients are struggling to ask, much less have the answers to. If a major fast food chain has hired you to help with "AI strategy," and you see their HR rushing to hire deep learning talent, it is your job to ask, "What problems with deep learning are you trying to solve?" If you cannot get a clear answer, you want to encourage management to step back and evaluate what real problems they actually are facing as an industry. Are they having staff-scheduling inefficiency? Well, they do not need deep learning. They need linear programming! This may seem basic to some readers, but a lot of management nowadays struggles to make these distinctions. More than once, I have met vendors and consultants who brand their linear programming solutions as AI, which can then semantically be conflated with deep learning.

What to Watch Out For in Data Science Jobs

To understand the data science job market, it might be good to draw a comparison to a profound piece of American television.

In 2010, there was a US TV series *Better Off Ted*. One episode, titled "Jabberwocky" (season 1, episode 12), captured something profound about corporate buzzwords. In the show Ted, the lead character, makes up a fictitious "Jabberwocky" project at his company to hide funding. With humorous results, his manager, the CEO, and ultimately the entire company began to "work" on the "Jabberwocky" project without even knowing what it is. It continues to escalate and thousands of employees are pretending to work on "Jabberwocky" while no one stops to ask what they are actually working on. The reason: no one wants to admit they are out of the loop and ignorant of something important.

The *Jabberwocky effect* is an anecdotal theory that an industry or organization can perpetuate a buzzword/project even if no one has satisfactorily defined what it is. Organizations can cyclically fall victim to this behavior, allowing terms to circulate without definition, and group behavior permits the ambiguity. Common examples may include blockchain, artificial intelligence, data science, big data, Bitcoin, internet of things, quantum computing, NFTs, being "data-driven," cloud computing, and "digital disruption." Even tangible, high-profile, and specific projects can become mysterious buzzwords understood by a few but talked about by many.

To stop the Jabberwocky effect, you have to be a catalyst for productive dialogue. Be curious about the methods and means (not just the qualities or outcomes) of a project or initiative. When it comes to a role, is the company hiring you to work on "Jabberwocky" or practical and specific projects? Have they fallen victim to buzzwords and hiring you because of FOMO (fear of missing out)? Or do they actually have specific and functional needs they are hiring you for? Making this discernment can spell the difference between a good fit that cruises smoothly or a frustrating speed bump in your career.

With that backdrop, let's now consider a few things to watch out for in data science jobs, starting with role definition.

Role Definition

Let's say you are hired as a data scientist. The interview went great. You asked questions about the role and got straight answers. You are offered a job, and most important, you should know what projects you will be working on.

You always want to go into a role that is clearly defined and has tangible objectives. There is no guessing what you are supposed to be working on. Even better, you should have leadership with a clear vision that understands what the business needs. You become the executor of clearly defined objectives and know your customer.

Conversely, if you were hired into a role because the department wants to be "data-driven" or have a competitive edge in "data science," this is a red flag. Chances are you will be burdened with finding problems to solve and selling any low-hanging fruit you find. When you ask for strategic guidance, you are told to apply "machine learning" to the business. But of course, when all you have is a hammer, everything starts to look like a nail. Data science teams feel pressured to have a solution (e.g., machine learning) before they even have an objective or problem to solve. Once a problem is found, getting stakeholder buy-in and aligning resources are found to be difficult, and focus starts to bounce from one low-hanging fruit to another.

The problem here is you were hired for a role based on a buzzword, not a function. Poor role definition tends to propagate to other problems discussed next. Let's move on to organizational focus.

Organizational Focus and Buy-In

Another factor to watch out for is how aligned the organization is on specific objectives and whether all parties are bought in.

Since the data science boom, many organizations restructured to have a central data science team. The executives' vision is to have the data science team rove, advise, and help other departments become data-driven and employ innovative techniques like machine learning. They may also be tasked with breaking down data silos between the departments. While this sounds like a good idea on paper, many organizations discover this is full of challenges.

Here's why: management creates a data science team but there's no clear objective. Therefore, this team is tasked with looking for problems to solve rather than empowered to solve known problems. As stated, this is why data science teams are notorious for having a solution (e.g., machine learning) before they have an objective. They are especially ill-equipped to be a driving force for breaking down data silos, as this is completely outside their lane of expertise.

Breaking Down Data Silos Is an IT Job!

It is misguided to use a data science team to "break down data silos" in an organization. Data silos are often due to a lack of data-warehousing infrastructure, and departments are housing their data in spreadsheets and secret databases rather than a centralized and supported database.

If data silos are deemed a problem, you need servers, cloud instances, certified database administrators, security protocols, and an IT task force to put all this together. This is not something a data science team is equipped to do, as they likely do not have the needed expertise, budget, and organizational authority to pull this off except in very small companies.

Once a problem is found, getting stakeholder buy-in and aligning resources are difficult. If an opportunity is found, strong leadership is needed to do the following:

- Have a clearly defined objective and road map
- Obtain budget to collect data and support the infrastructure
- Attain data access and negotiate data ownership
- Include stakeholder buy-in and domain knowledge
- Budget time and meetings from stakeholders

These requirements are much harder to do *after* a data science team has been hired, rather than before, because the data science team's roles are being scoped and budgeted reactively. If higher leadership has not aligned resources and buy-in from all necessary parties, the data science project will not succeed. This is why there is no shortage of articles blaming organizations for not being ready for data science, from *Harvard Business Review* (*https://oreil.ly/IlicW*) to *MIT Sloan Review* (*https://oreil.ly/U9C9F*).

It is better to work on a data science team that organizationally is in the same department as its customer. Information, budgets, and communication are shared more freely and cohesively. That way, there is less tension, as interdepartmental politics are mitigated by putting everyone on the same team rather than in political competition.

Data Access Is Political

It is no secret organizations are protective of their data, but this is not just due to security or distrust concerns. Data itself is a highly political asset, and many people are reluctant to provide access even to their own colleagues. Even departments inside the same organization will not share data with one another for this reason: they do not want others doing their job, much less doing it incorrectly. It may require *their* full-time expertise to interpret the data, and it requires *their* domain knowledge. After all, their data is their business! And if you are asking for access to *their* data, you are asking to get into *their* business.

On top of that, data scientists can overestimate their ability to interpret foreign datasets and the domain expertise needed to use them. To overcome this hurdle, you have to develop trust and buy-in with each partner with expertise, negotiate a knowledge transfer, and if needed, give them a significant role in the project.

Adequate Resources

Another risk to watch out for is not getting adequate resources to do your job. It is difficult being thrown into a role and not having what you need. Of course, being scrappy and resourceful is an invaluable trait. But even the scrappiest software engineer/data scientist rock star can quickly find themselves in over their head. Sometimes you need things that cost money and your employer cannot budget for it.

Let's say you need a database to perform forecasting work. You have a poor connection to a third-party database, with frequent downtime and disconnects. The last thing you want to hear is to "make it work" but that is your situation. You contemplate replicating the database locally but to do that, you need to store 40 GB a day, and therefore you need a server or cloud instance. Now you are clearly in over your head, a data scientist becoming an IT department without an IT budget!

In these situations, you have to contemplate how you can cut corners without hurting your project. Can you hold only the rolling latest data and delete the rest? Can you create some error-handling Python script that reconnects when a disconnect happens, while breaking up the data in batches so it picks up from the last successful batch?

If this problem/solution sounds specific, yes, I did have to do this, and yes, it did work! It is satisfying to come up with workaround solutions and streamline a process without incurring more costs. But inevitably, for many data projects you may need data pipelines, servers, clusters, GPU-based workstations, and other computational resources that a desktop computer cannot offer. In other words, these things can cost money and your organization may be unable to budget for them.

Where's the Mathematical Modeling?

If you are wondering how you were hired to do regression, statistics, machine learning, and other applied mathematics only to find yourself doing rogue IT work, this is not uncommon in the current corporate climate.

You are working with data, however, and that implicitly can lead to IT-like work. The important thing is to make sure your skills still match the job and the needed outcome. We will address this throughout the rest of this chapter.

Reasonable Objectives

This is a big one to watch out for. In a landscape full of hype and moonshot promises, it is easy to encounter unreasonable objectives.

There are situations where a manager hires a data scientist and expects them to waltz in effortlessly and add exponential value to the organization. This can certainly happen if the organization is still doing manual work and opportunities to automate are everywhere. For example, if the organization is doing all its work in spreadsheets and forecasts are done by pure guesses, this is a great opportunity for a data science professional to streamline processes into a database and make strides with even simple regression models.

On the other hand, if the organization hires a data scientist to implement machine learning into their software specifically recognizing objects in images, this is much harder. A well-informed data scientist has to explain to management this is an endeavor that will cost hundreds of thousands of dollars at least! Not only do pictures have to be gathered, but manual labor has to be hired to label the objects in the images (*https://oreil.ly/ov7S5*). And that is just collecting the data!

It is common for a data scientist to spend their first 18 months explaining to management why they have not delivered, because they are still trying to gather and clean data, which is 95% of machine learning efforts. Management can become disillusioned with this because they fell victim to a popular narrative that machine learning and AI would do away with manual processes, only to discover they traded one set of manual processes for another: the procurement of labeled data.

So be wary of environments that set unreasonable objectives, and find diplomatic ways to manage expectations with management especially when others promised them an "EASY button." There are a lot of claims from otherwise reputable business journals and high-dollar management consultancies that super-intelligent AI is around the corner. Managers who lack technical expertise can fall victim to this hyped narrative.

Cui Bono?

The Latin expression cui bono means "who benefits?" It is a good question to ask when you find yourself trying to make sense of behaviors when the Jabberwocky effect is in full swing. When the media promotes stories about artificial intelligence, who benefits from that? Regardless of your answer, the media also benefits from the clicks and ad revenue. High-dollar management consultancies create more billable hours around "AI strategy." A chip manufacturer can promote deep learning to sell more graphics cards, and cloud platforms can sell more data storage and CPU time for machine learning projects.

What do all of these parties have in common? It is not just that they are using AI as a means to sell their products, but they have no long-term stake in their customers' success. They are selling units, not a project outcome, much like selling shovels during a gold rush.

However, I'm not saying these media and vendor motivations are unethical. It is their employees' job to make money for their company and provide for their families. The claims that promote their product may even be legitimate and achievable. However, it cannot be dismissed that once a claim is promoted, it is difficult to walk it back, even when it is realized to be unattainable. Many businesses might pivot and redirect their efforts rather than admit their claims did not pan out. So just be aware of this dynamic and always ask "cui bono?"

Competing with Existing Systems

This caution might fall under "reasonable objectives," but I think this situation is common enough that it warrants its own category. A subtle but problematic type of role is one that competes with an existing system that is not actually broken. These situations may arise in work environments that lack things to do and need to look busy.

Your employer contracted a vendor to install a sales-forecasting system several years ago. Your manager now asks you to improve the forecasting system by increasing accuracy by 1%.

Do you see the statistical problem here? If you read Chapter 3, 1% should not feel statistically significant, and randomness can easily give you that 1% without any effort on your part. Conversely, randomness can swing the opposite direction and market forces out of your control can negate any effort you implement. A bad sales quarter and factors outside your control (such as competitors moving into your company's market) decrease revenue by -3% rather than the 1% you inevitably p-hacked.

The overarching problem here, besides the redundancy of work, is the outcome is not in your sphere of influence. This may become a less than optimal situation. It's one thing if the existing system you are competing with is broken and rudimentary, or done completely manually without automation. But competing with a system that is not broken is setting yourself up for a bad time. If possible, run for the hills when you see this kind of project.

"What Would You Say...You Do Here?"

Can a data scientist be hired into a role that produces no value, despite good work and diligent efforts? Yes. Factors outside one's control can nullify even the best work, and it is important to keep a sharp eye out for this.

The 1999 Mike Judge comedy film *Office Space* became a cult classic for many corporate workers in the United States. In the movie, IT worker and protagonist Peter Gibbons mentally checks out of a coding job reporting to eight different managers. When downsizing consultants ask what he does during his day, he answers honestly that he does "about 15 minutes of real, actual work." I won't spoil the rest of the movie for uninitiated readers, but like any great comedy, the outcome is probably not what you expect.

To elaborate on the previous example, replacing a system that is not broken is what late anthropologist David Graeber would describe as a *bullshit job* (pardon my French). According to Graeber, a bullshit job is paid employment that is so completely pointless, unnecessary, or pernicious that even the employee cannot justify its existence but has to pretend otherwise. In his book *Bullshit Jobs* and viral 2013 article (*https://www.strike.coop/bullshit-jobs*), Graeber theorizes these jobs are becoming so commonplace that it is taking a psychological toll on the workforce and economy.

While Graeber's work is riddled with self-selection bias and anecdotal findings, and the lack of empirical evidence is an arguable target for criticism, it is hard-pressed to claim these bullshit jobs do not exist. In Graeber's defense, this is empirically hard to measure and few workers will answer a poll honestly for fear of jeopardizing their own careers.

Are data science careers safe from these problems? Cassie Kozykrov, the Chief Decision Scientist at Google, shares a strange anecdote (*https://oreil.ly/fwPKn*) that helps answer this question:

> Several years ago, an engineering director friend who works in tech was bemoaning his useless data scientists. "I think you might be hiring data scientists the way a drug lord buys a tiger for his backyard," I told him. "You don't know what you want with the tiger, but all the other drug lords have one."

Yikes. Is it possible management hires data scientists to increase an organization's perceived credibility and corporate standing? If you find yourself in a job not designed to create value, think strategically how you can influence positive change. Can you create opportunity rather than waiting on others to provide it? Can you meaningfully take ownership of initiatives you identify and therefore grow your career? If this is not possible, empower yourself to pursue better opportunities.

A Role Is Not What You Expected

When you start a role and find out it is not what you expected, what do you do? For example, you were told your role would be statistics and machine learning but you find yourself doing IT-like work instead, as the data foundation of the organization is simply not developed enough to do machine learning.

You might be able to make lemonade. You can certainly accept your data scientist role turning into an IT role, and maybe get some database and programming skills in the process. You might even become indispensable as the resident SQL expert or technical guru, and it can be professionally comfortable. As you streamline your business's data operations and workflows, you are setting your business up to be better prepared for more sophisticated applications down the road. As your operations run smoothly, you can allocate time to learn and grow professionally in what interests you.

On the other hand, if you were expecting to do statistical analysis and machine learning, and instead find yourself debugging broken spreadsheets, Microsoft Access, and VBA macros, you will probably feel disappointment. In this situation, at least become an advocate for change. Push to modernize the tooling, advocating using Python and a modern database platform, like MySQL or even SQLite. If you can achieve that, then at least you will be on a platform that allows more innovation, and be that much closer to applying concepts in this book. This will also benefit the organization, as support and flexibility of tools will increase, and Python talent is more readily available than dated technologies like Microsoft Access and VBA.

What Is Shadow IT?

Shadow information technology (shadow IT) (*https://oreil.ly/9ZDb8*) is a term describing office workers who create systems outside their IT department. These systems can include databases, scripts, and processes as well as vendor and employee-made software without the IT department's involvement.

Shadow IT used to be frowned on in organizations, as it is unregulated and operating outside the IT department's scope of control. When finance, marketing, and other non-IT departments decide to hand-spin their own rogue IT operations, it certainly can create hidden costs to the organization in inefficiency and security concerns. Nasty politics can ensue when IT departments and non-IT departments clash, accusing each other of not staying in their lane or simply co-opting roles for job security.

However, one benefit of the data science movement is it has made shadow IT more accepted as a necessity for innovation. Non-IT folks can prototype and experiment with creating datasets, Python scripts, and regression tools. The IT department can then pick up these innovations and support them formally as they mature. This also allows the business to become more nimble. A business rule change can be a quick edit to a Python script or in-house database, rather than filing a ticket with the IT help desk. Whether that change should be subject to rigorous testing and red tape is certainly a cost trade-off for responsiveness.

Overall, if you find yourself in a shadow IT role (which is likely), make sure you understand the risks and play nice with the IT department. If you are successful, it can be empowering and rewarding to support your business this way. If you sense possible conflict with your IT department, be up front with your leadership and let them know. If you can truthfully say your work is "prototyping" and "exploration," then your leadership can argue it is outside the IT purview. Never go rogue without your leadership's support, though, and let them deal with the interdepartmental politics.

Does Your Dream Job Not Exist?

While you can always walk away from a role, always be sure to assess how realistic your expectations are, too. Are the cutting-edge technologies you are pursuing perhaps too cutting-edge?

Take natural language processing for example. You want to build chatbots using deep learning. However, the number of actual jobs that do this are scarce, as most companies do not have practical need for chatbots. Why? Chatbots are simply not there yet. While companies like OpenAI have interesting research like GPT-3 (*https:// openai.com/blog/gpt-3-apps*), that is mostly what it amounts to: interesting research. GPT-3 in the end is a probability-based pattern recognizer chaining words together, and therefore it has no common sense. There is research demonstrating this, including some by Gary Marcus at New York University (*https://oreil.ly/fxakC*).

This means creating chatbots with broader applications is an open problem and has yet to be a value proposition for a great majority of businesses. If natural language processing is something you really want to pursue, and you are finding a disconnect with career opportunities, your best bet might be to go into the academic world and do research. While there are companies like Alphabet that do academic-like research, many of those employees came from academia.

So again, keep your expectations realistic as you navigate the job market. If your expectations exceed what the job market can provide, strongly consider the academic route. You also should consider this route when the type of job you want frequently requires a PhD or specific academic credentials, and this becomes a barrier to doing your dream job.

Where Do I Go Now?

Now that we covered the data science landscape, where do we go from here? And what is the future of data science?

First, consider the burdens of having a data scientist job title. It carries an implicit requirement to have knowledge without boundaries, mostly due to the lack of standardized definition with a restricted scope. If we learned anything from the last 10 years watching the data science movement, definition matters. A data scientist is evolving to be a software engineer with proficiency in statistics, optimization, and machine learning. A data scientist may not even have that data scientist title anymore. While this is a much broader umbrella of requirements than when the data scientist was declared the "sexiest job of the 21st century," it is becoming necessary to have these skills.

Another option is to specialize in favor of more focused titles, and this has been happening increasingly for the past few years. Roles like computer vision engineer, data engineer, data analyst, researcher, operations research analyst, and advisor/consultant are making a comeback. We are seeing less of the data scientist role, and this trend is likely to continue in the next 10 years, primarily due to role specialization. It is certainly an option to follow this trend.

It is critical to note that the labor market has changed drastically, and this is why you need the competitive edges listed in this chapter. While data scientists were seen as unicorns in 2014 commanding six-figure salaries, nowadays a given data scientist job at any company can easily receive hundreds or thousands of applications while only offering a five-figure salary. Data science degrees and bootcamps have created an enormous boom of supply in data science professionals. Therefore, it is competitive landing jobs marketed as a data scientist or data science in general. This is why going after roles like analyst, operations research, and software developer is not necessarily a bad idea! Vicki Boykis, a machine learning engineer at Tumblr, probably says it best on her blog article "Data Science is Different Now" (*https://oreil.ly/vm8Vp*):

> Remember that the ultimate goal...is to beat the hordes doing data science degrees, bootcamps, and working through tutorials.
>
> You want to get your foot in the door, *get a data-adjacent position*, and move towards whatever your dream job is, while finding out as much as you can about the tech industry in general.
>
> Don't get paralysis by analysis. Pick a small piece of something and start there. Do something small. Learn something small, build something small. Tell other people. Remember that your first job in data science will probably not be as a data scientist.

Conclusion

This was a different chapter from the rest in this book, but it is important if you want to navigate the data science job landscape and effectively apply the knowledge in this book. It might be disconcerting to learn statistical tools and machine learning only to find much of the job landscape will pivot you into other work. When this happens, take the opportunity to continue learning and gaining skills. When you integrate your essential math knowledge with a programming and software engineering proficiency, your value will increase tenfold just from having an understanding of the IT and data science gap.

Remember to eschew the hype in favor of practical solutions, and do not get so caught up in the technical purview that you become blindsided by market forces. Understand management and leadership motivations, as well as the motivations of people in general. Understand *why* things work, not just *how* they work. Be curious *why* a technique or tool solves a problem, not just *how* the technical aspects of it operate.

Learn not for the sake of it, but to develop capabilities and pair the right tool to the right problem. One of the most effective ways to learn is to pick a problem (not a tool!) you find interesting. Pulling that thread leads to another thing to be curious about, and then another, and then another. You have a goal in mind, so continue going down the right rabbit holes and know when to pull yourself out of others. It is absolutely rewarding to take this approach, and it is amazing how much expertise you can gain in a short amount of time.

Supplemental Topics

Using LaTeX Rendering with SymPy

As you get more comfortable with mathematical notation, it can be helpful to take your SymPy expressions and display them in mathematical notation.

The quickest way to do this is to use the `latex()` function in SymPy on your expression and then copy the result to a LaTeX math viewer.

Example A-1 is an example that takes a simple expression and turns it into a LaTeX string. Of course we can take the results of derivatives, integrals, and other SymPy operations and render those as LaTeX too. But let's keep the example straightforward.

Example A-1. Using SymPy to convert an expression into LaTeX

```
from sympy import *

x,y = symbols('x y')

z = x**2 / sqrt(2*y**3 - 1)

print(latex(z))

# prints
# \frac{x^{2}}{\sqrt{2 y^{3} - 1}}
```

This `\frac{x^{2}}{\sqrt{2 y^{3} - 1}}` string is formatted mathlatex, and there are a variety of tools and document formats that can be adapted to support it. But to simply render the mathlatex, go to a LaTeX equation editor. Here are two different ones I use online:

- Lagrida LaTeX Equation Editor (*https://latexeditor.lagrida.com*)
- CodeCogs Equation Editor (*https://latex.codecogs.com*)

In Figure A-1 I use Lagrida's LaTeX editor to render the mathematical expression.

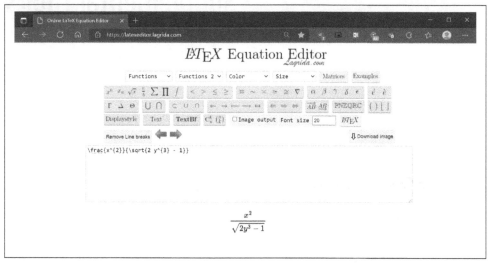

Figure A-1. Using a math editor to view the SymPy LaTeX output

If you want to save the copy/paste step, you can append the LaTeX directly as an argument to the CodeCogs LaTeX editor URL as shown in Example A-2, and it will show the rendered math equation in your browser.

Example A-2. Open a mathlatex rendering using CodeCogs

```
import webbrowser
from sympy import *

x,y = symbols('x y')

z = x**2 / sqrt(2*y**3 - 1)

webbrowser.open("https://latex.codecogs.com/png.image?\dpi{200}" + latex(z))
```

If you use Jupyter, you can also use plugins to render mathlatex (*https://oreil.ly/mWYf7*).

Binomial Distribution from Scratch

If you want to implement a binomial distribution from scratch, here are all the parts you need in Example A-3.

Example A-3. Building a binomial distribution from scratch

```
# Factorials multiply consecutive descending integers down to 1
# EXAMPLE: 5! = 5 * 4 * 3 * 2 * 1
def factorial(n: int):
    f = 1
    for i in range(n):
        f *= (i + 1)
    return f

# Generates the coefficient needed for the binomial distribution
def binomial_coefficient(n: int, k: int):
    return factorial(n) / (factorial(k) * factorial(n - k))


# Binomial distribution calculates the probability of k events out of n trials
# given the p probability of k occurring
def binomial_distribution(k: int, n: int, p: float):
    return binomial_coefficient(n, k) * (p ** k) * (1.0 - p) ** (n - k)

# 10 trials where each has 90% success probability
n = 10
p = 0.9

for k in range(n + 1):
    probability = binomial_distribution(k, n, p)
    print("{0} - {1}".format(k, probability))
```

Using the `factorial()` and the `binomial_coefficient()`, we can build a binomial distribution function from scratch. The factorial function multiplies a consecutive range of integers from 1 to n. For example, a factorial of 5! would be $1*2*3*4*5 = 120$.

The binomial coefficient function allows us to select k outcomes from n possibilities with no regard for ordering. If you have $k = 2$ and $n = 3$, that would yield sets (1,2) and (1,2,3), respectively. Between those two sets, the possible distinct combinations would be (1,3), (1,2), and (2,3). That is three combinations so that would be a binomial coefficient of 3. Of course, using the `binomial_coefficient()` function we can avoid all that permutation work by using factorials and multiplication instead.

When implementing `binomial_distribution()`, notice how we take the binomial coefficient and multiply it by the probability of success p occurring k times (hence the exponent). We then multiply it by the opposite case: the probability of failure

`1.0 - p` occurring `n - k` times. This allows us to account for the probability p of an event occurring versus not occurring across several trials.

Beta Distribution from Scratch

If you are curious how to build a beta distribution from scratch, you will need to reuse the `factorial()` function we used for the binomial distribution as well as the `approximate_integral()` function we built in Chapter 2.

Just like we did in Chapter 1, we pack rectangles under the curve for the range we are interested in as shown in Figure A-2.

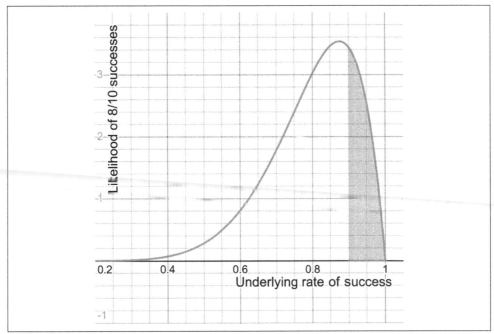

Figure A-2. Packing rectangles under the curve to find the area/probability

This is using just six rectangles; we will get better accuracy if we were to use more rectangles. Let's implement the `beta_distribution()` from scratch and integrate 1,000 rectangles between 0.9 and 1.0 as shown in Example A-4.

Example A-4. Beta distribution from scratch

```
# Factorials multiply consecutive descending integers down to 1
# EXAMPLE: 5! = 5 * 4 * 3 * 2 * 1
def factorial(n: int):
    f = 1
    for i in range(n):
```

```
        f *= (i + 1)
    return f

def approximate_integral(a, b, n, f):
    delta_x = (b - a) / n
    total_sum = 0

    for i in range(1, n + 1):
        midpoint = 0.5 * (2 * a + delta_x * (2 * i - 1))
        total_sum += f(midpoint)

    return total_sum * delta_x

def beta_distribution(x: float, alpha: float, beta: float) -> float:
    if x < 0.0 or x > 1.0:
        raise ValueError("x must be between 0.0 and 1.0")

    numerator = x ** (alpha - 1.0) * (1.0 - x) ** (beta - 1.0)
    denominator = (1.0 * factorial(alpha - 1) * factorial(beta - 1)) / \
            (1.0 * factorial(alpha + beta - 1))

    return numerator / denominator


greater_than_90 = approximate_integral(a=.90, b=1.0, n=1000,
    f=lambda x: beta_distribution(x, 8, 2))
less_than_90 = 1.0 - greater_than_90

print("GREATER THAN 90%: {}, LESS THAN 90%: {}".format(greater_than_90,
    less_than_90))
```

You will notice with the beta_distribution() function, we provide a given probability x, an alpha value quantifying successes, and a beta value quantifying failures. The function will return how likely we are to observe a given likelihood x. But again, to get a probability of observing probability x we need to find an area within a range of x values.

Thankfully, we have our approximate_integral() function defined and ready to go from Chapter 2. We can calculate the probability that the success rate is greater than 90% as well as less than 90%, as shown in the last few lines.

Deriving Bayes' Theorem

If you want to understand why Bayes' Theorem works rather than take my word for it, let's do a thought experiment. Let's say I have a population of 100,000 people. Multiply it with our given probabilities to get the count of people who drink coffee and the count of people who have cancer:

$N = 100,000$

$P(\text{Coffee Drinker}) = .65$

$P(\text{Cancer}) = .005$

Coffee Drinkers = 65,000

Cancer Patients = 500

We have 65,000 coffee drinkers and 500 cancer patients. Now of those 500 cancer patients, how many are coffee drinkers? We were provided with a conditional probability $P(\text{Coffee}|\text{Cancer})$ we can multiply against those 500 people, which should give us 425 cancer patients who drink coffee:

$P(\text{Coffee Drinker}|\text{Cancer}) = .85$

Coffee Drinkers with Cancer = $500 \times .85 = 425$

Now what is the percentage of coffee drinkers who have cancer? What two numbers do we divide? We already have the number of people who drink coffee *and* have cancer. Therefore, we proportion that against the total number of coffee drinkers:

$$P(\text{Cancer}|\text{Coffee Drinker}) = \frac{\text{Coffee Drinkers with Cancer}}{\text{Coffee Drinkers}}$$

$$P(\text{Cancer}|\text{Coffee Drinker}) = \frac{425}{65,000}$$

$$P(\text{Cancer}|\text{Coffee Drinker}) = 0.006538$$

Hold on a minute, did we just flip our conditional probability? Yes we did! We started with $P(\text{Coffee Drinker}|\text{Cancer})$ and ended up with $P(\text{Cancer}|\text{Coffee Drinker})$. By taking two subsets of the population (65,000 coffee drinkers and 500 cancer patients), and then applying a joint probability using the conditional probability we had, we ended up with 425 people in our population who both drink coffee and have cancer. We then divide that by the number of coffee drinkers to get the probability of cancer given one's a coffee drinker.

But where is Bayes' Theorem in this? Let's focus on the $P(\text{Cancer}|\text{Coffee Drinker})$ expression and expand it with all the expressions we previously calculated:

$$P(\text{Cancer}|\text{Coffee Drinker}) = \frac{100,000 \times P(\text{Cancer}) \times P(\text{Coffee Drinker}|\text{Cancer})}{100,000 \times P(\text{Coffee Drinker})}$$

Notice the population N of 100,000 exists in both the numerator and denominator so it cancels out. Does this look familiar now?

$$P(\text{Cancer}|\text{Coffee Drinker}) = \frac{P(\text{Cancer}) \times P(\text{Coffee Drinker}|\text{Cancer})}{P(\text{Coffee Drinker})}$$

Sure enough, this should match Bayes' Theorem!

$$P(A|B) = \frac{P(B|A) * P(B)}{P(A)}$$

$$P(\text{Cancer}|\text{Coffee Drinker}) = \frac{P(\text{Cancer}) \times P(\text{Coffee Drinker}|\text{Cancer})}{P(\text{Coffee Drinker})}$$

So if you get confused by Bayes' Theorem or struggle with the intuition behind it, try taking subsets of a fixed population based on the provided probabilities. You can then trace your way to flip a conditional probability.

CDF and Inverse CDF from Scratch

To calculate areas for the normal distribution, we can of course use the rectangle-packing method we learned in Chapter 1 and applied to the beta distribution earlier in the appendix. It doesn't require the cumulative density function (CDF) but simply packed rectangles under the probability density function (PDF). Using this method, we can find the probability a golden retriever weighs between 61 and 62 pounds as shown in Example A-5, using 1,000 packed rectangles against the normal PDF.

Example A-5. The normal distribution function in Python

```python
import math


def normal_pdf(x: float, mean: float, std_dev: float) -> float:
    return (1.0 / (2.0 * math.pi * std_dev ** 2) ** 0.5) *
        math.exp(-1.0 * ((x - mean) ** 2 / (2.0 * std_dev ** 2)))


def approximate_integral(a, b, n, f):
    delta_x = (b - a) / n
    total_sum = 0

    for i in range(1, n + 1):
        midpoint = 0.5 * (2 * a + delta_x * (2 * i - 1))
        total_sum += f(midpoint)

    return total_sum * delta_x
```

```
p_between_61_and_62 = approximate_integral(a=61, b=62, n=7,
    f= lambda x: normal_pdf(x,64.43,2.99))

print(p_between_61_and_62) # 0.0825344984983386
```

That will give us about 8.25% probability a golden retriever weighs between 61 and 62 pounds. If we wanted to leverage a CDF that is already integrated for us and does not require any rectangle packing, we can declare it from scratch as shown in Example A-6.

Example A-6. Using the inverse CDF (called `ppf()`) in Python

```
import math

def normal_cdf(x: float, mean: float, std_dev: float) -> float:
    return (1 + math.erf((x - mean) / math.sqrt(2) / std_dev)) / 2

mean = 64.43
std_dev = 2.99

x = normal_cdf(66, mean, std_dev) - normal_cdf(62, mean, std_dev)

print(x)  # prints 0.49204501470628936
```

The `math.erf()` is known as the error function and is often used to compute cumulative distributions. Finally, to do the inverse CDF from scratch you will need to use the inverse of the `erf()` function called `erfinv()`. Example A-7 calculates one thousand randomly generated golden retriever weights using an inverse CDF coded from scratch.

Example A-7. Generating random golden retriever weights

```
import random
from scipy.special import erfinv

def inv_normal_cdf(p: float, mean: float, std_dev: float):
    return mean + (std_dev * (2.0 ** 0.5) * erfinv((2.0 * p) - 1.0))

mean = 64.43
std_dev = 2.99

for i in range(0,1000):
    random_p = random.uniform(0.0, 1.0)
    print(inv_normal_cdf(random_p, mean, std_dev))
```

Use e to Predict Event Probability Over Time

Let's look at one more use case for e that you might find useful. Let's say you are a manufacturer of propane tanks. Obviously, you do not want the tank to leak or else that could create hazards, particularly around open flames and sparks. Testing a new tank design, your engineer reports that there is a 5% chance in a given year that it will leak.

You know this is already an unacceptably high number, but you want to know how this probability compounds over time. You now ask yourself, "What is the probability of a leak happening within 2 years? 5 years? 10 years?" The more time that is exposed, would not the probability of seeing the tank leak only get higher? Euler's number can come to the rescue again!

$$P_{leak} = 1.0 - e^{-\lambda T}$$

This function models the probability of an event over time, or in this case the tank leaking after T time. e again is Euler's number, lambda λ is the failure rate across each unit of time (each year), and T is the amount of time gone by (number of years).

If we graph this function where T is our x-axis, the probability of a leak is our y-axis, and $\lambda = .05$, Figure A-3 shows what we get.

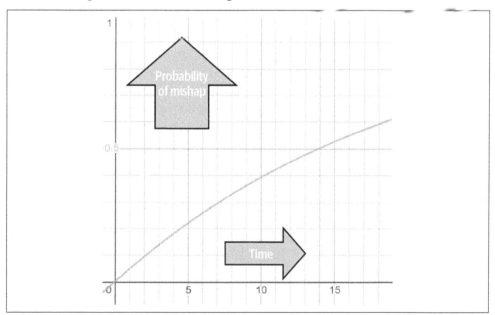

Figure A-3. Predicting the probability of a leak over time

Here is how we model this function in Python for $\lambda = .05$ and $T = 5$ years in Example A-8.

Example A-8. Code for predicting the probability of a leak over time

```
from math import exp

# Probability of leak in one year
p_leak = .05

# number of years
t = 5

# Probability of leak within five years
# 0.22119921692859512
p_leak_5_years = 1.0 - exp(-p_leak * t)

print("PROBABILITY OF LEAK WITHIN 5 YEARS: {}".format(p_leak_5_years))
```

The probability of a tank failure after 2 years is about 9.5%, 5 years is about 22.1%, and 10 years 39.3%. The more time that passes, the more likely the tank will leak. We can generalize this formula to predict any event with a probability in a given period and see how that probability shifts over different periods of time.

Hill Climbing and Linear Regression

If you find the calculus overwhelming in building machine learning from scratch, you can try a more brute-force method. Let's try a *hill climbing* algorithm, where we randomly adjust m and b by adding random values for a number of iterations. These random values will be positive or negative (which will make the addition operation effectively subtraction), and we will only keep adjustments that improve our sum of squares.

But do we just generate any random number as the adjustment? We will want to prefer smaller moves but occasionally we might allow larger moves. This way, we have mostly fine adjustments, but occasionally we will make big jumps if they are needed. The best tool to do this is a standard normal distribution, with a mean of 0 and a standard deviation of 1. Recall from Chapter 3 that a standard normal distribution will have a high density of values near 0, and the farther the value is away from 0 (in both the negative and positive direction), the less likely the value becomes as shown in Figure A-4.

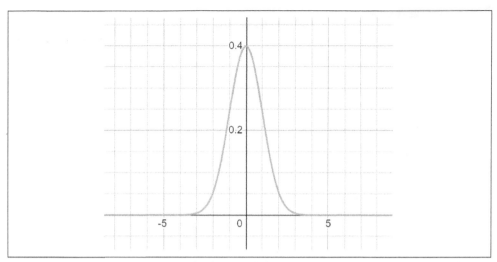

Figure A-4. Most values in a standard normal distribution are small and near 0, while larger values are less frequent on the tails

Circling back to the linear regression, we will start m and b at 0 or some other starting values. Then for 150,000 iterations in a for loop, we will randomly adjust m and b by adding values sampled from the standard normal distribution. If a random adjustment improves/lessens the sum of squares, we keep it. But if the sum of squares increases, we undo that random adjustment. In other words, we only keep adjustments that improve the sum of squares. Let's take a look in Example A-9.

Example A-9. Using hill climbing for linear regression

```
from numpy.random import normal
import pandas as pd

# Import points from CSV
points = [p for p in pd.read_csv("https://bit.ly/2KF29Bd").itertuples()]

# Building the model
m = 0.0
b = 0.0

# The number of iterations to perform
iterations = 150000

# Number of points
n = float(len(points))

# Initialize with a really large loss
# that we know will get replaced
best_loss = 10000000000000.0
```

```
for i in range(iterations):

    # Randomly adjust "m" and "b"
    m_adjust = normal(0,1)
    b_adjust = normal(0,1)

    m += m_adjust
    b += b_adjust

    # Calculate loss, which is total sum squared error
    new_loss = 0.0
    for p in points:
        new_loss += (p.y - (m * p.x + b)) ** 2

    # If loss has improved, keep new values. Otherwise revert.
    if new_loss < best_loss:
        print("y = {0}x + {1}".format(m, b))
        best_loss = new_loss
    else:
        m -= m_adjust
        b -= b_adjust

print("y = {0}x + {1}".format(m, b))
```

You will see the progress of the algorithm, but ultimately you should get a fitted function of approximately y = 1.9393722016562853x + 4.731834051245578, give or take. Let's validate this answer. When I used Excel or Desmos to perform a linear regression, Desmos gave me y = 1.93939x + 4.73333. Not bad! I got pretty close!

Why did we need one million iterations? Through experimentation, I found this with enough iterations where the solution was not really improving much anymore and converged closely to the optimal values for m and b to minimize the sum of squares. You will find many machine learning libraries and algorithms have a parameter for the number of iterations to perform, and it does exactly this. You need to have enough so it converges on the right answer approximately, but not so much that it wastes computation time when it has already found an acceptable solution.

One other question you may have is why I started the best_loss at an extremely large number. I did this to initialize the best loss with a value I know will be overwritten once the search starts, and it will then be compared to the new loss of each iteration to see if it results in an improvement. I also could have used positive infinity float('inf') instead of a very large number.

Hill Climbing and Logistic Regression

Just like in the previous example with linear regression, we can also apply hill climbing to logistic regression. Again, use this technique if you find the calculus and partial derivatives to be too much at once.

The hill climbing methodology is identical: adjust m and b with random values from a normal distribution. However we do have a different objective function, the maximum likelihood estimation, as discussed in Chapter 6. Therefore, we only take random adjustments that increase the likelihood estimation, and after enough iterations we should converge on a fitted logistic regression.

This is all demonstrated in Example A-10.

Example A-10. Using hill climbing for a simple logistic regression

```python
import math
import random

import numpy as np
import pandas as pd


# Desmos graph: https://www.desmos.com/calculator/6cb10atg3l

points = [p for p in pd.read_csv("https://tinyurl.com/y2cocoo/").itertuples()]

best_likelihood = -10_000_000
b0 = .01
b1 = .01


# calculate maximum likelihood

def predict_probability(x):
    p = 1.0 / (1.0001 + math.exp(-(b0 + b1 * x)))
    return p


for i in range(1_000_000):

    # Select b0 or b1 randomly, and adjust it randomly
    random_b = random.choice(range(2))

    random_adjust = np.random.normal()

    if random_b == 0:
        b0 += random_adjust
    elif random_b == 1:
        b1 += random_adjust
```

```
# Calculate total likelihood
true_estimates = sum(math.log(predict_probability(p.x)) \
    for p in points if p.y == 1.0)
false_estimates = sum(math.log(1.0 - predict_probability(p.x)) \
    for p in points if p.y == 0.0)

total_likelihood = true_estimates + false_estimates

# If likelihood improves, keep the random adjustment. Otherwise revert.
if best_likelihood < total_likelihood:
    best_likelihood = total_likelihood
elif random_b == 0:
    b0 -= random_adjust
elif random_b == 1:
    b1 -= random_adjust

print("1.0 / (1 + exp(-({0} + {1}*x))".format(b0, b1))
print("BEST LIKELIHOOD: {0}".format(math.exp(best_likelihood)))
```

Refer to Chapter 6 for more details on the maximum likelihood estimation, the logistic function, and the reason we use the log() function.

A Brief Intro to Linear Programming

A technique that every data science professional should be familiar with is *linear programming*, which solves a system of inequalities by adapting systems of equations with "slack variables." When you have variables that are discrete integers or binaries (0 or 1) in a linear programming system, it is known as *integer programming*. When linear continuous and integer variables are used, it is known as *mixed integer programming*.

While it is much more algorithm-driven than data-driven, linear programming and its variants can be used to solve a wide array of classic AI problems. If it sounds dubious to brand linear programming systems as AI, it is common practice by many vendors and companies as it increases the perceived value.

In practice, it is best to use the many available solver libraries to do linear programming for you, but resources at the end of this section will be provided on how to do it from scratch. For these examples we will use PuLP (*https://pypi.org/project/PuLP*), although Pyomo (*https://www.pyomo.org*) is an option as well. We will also use graphical intuition, although problems with more than three dimensions cannot be visualized easily.

Here's our example. You have two lines of products: the iPac and iPac Ultra. The iPac makes $200 profit while the iPac Ultra makes $300 profit.

However, the assembly line can work for only 20 hours, and it takes 1 hour to produce the iPac and 3 hours to produce an iPac Ultra.

Only 45 kits can be provided in a day, and an iPac requires 6 kits while iPac Ultra requires 2 kits.

Assuming all supply will be sold, how many of the iPac and iPac Ultra should we sell to maximize profit?

Let's first note that first constraint and break it down:

> ...the assembly line can work for only 20 hours, and it takes 1 hour to produce the iPac and 3 hours to produce an iPac Ultra.

We can express that as an inequality where x is the number of iPac units and y is the number of iPac Ultra units. Both must be positive and Figure A-5 shows we can graph accordingly.

$$x + 3y \leq 20 (x \geq 0, y \geq 0)$$

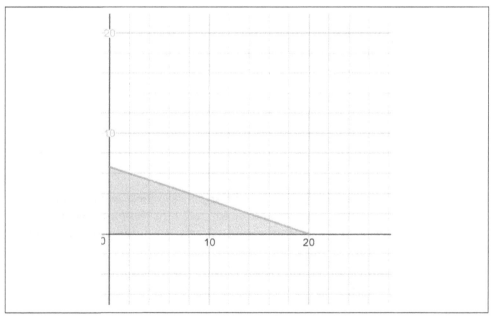

Figure A-5. Graphing the first constraint

Now let us look at the second constraint:

> Only 45 kits can be provided in a day, and an iPac requires 6 kits while iPac Ultra requires 2 kits.

We can also model and graph in Figure A-6 accordingly.

$$6x + 2y \leq 45 (x \geq 0, y \geq 0)$$

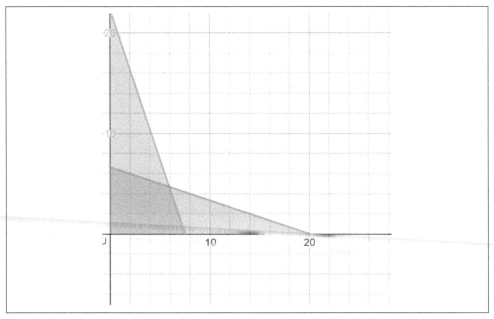

Figure A-6. Graphing the second constraint

Notice in Figure A-6 that we now have an overlap between these two constraints. Our solution is somewhere in that overlap and we will call it the *feasible region*. Finally, we are maximizing our profit Z, which is expressed next, given the profit amounts for the iPac and iPac Ultra, respectively.

$$Z = 200x + 300y$$

If we express this function as a line, we can increase Z as much as possible until the line is just no longer in the feasible region. We then note the x- and y-values as visualized in Figure A-7.

Desmos Graph of the Objective Function

If you need to see this visualized in a more interactive and animated fashion, check out this graph on Desmos (*https://oreil.ly/RQMBT*).

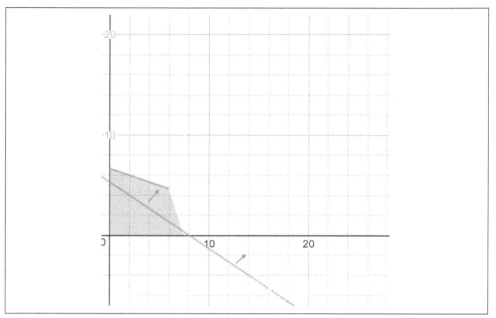

Figure A-7. Increasing our objective line until it no longer is in the feasible region

When that line "just touches" the feasible region as you increase profit Z as much as possible, you will land on a vertex, or a corner, of the feasible region. That vertex provides the x- and y-values that will maximize profit, as shown in Figure A-8.

While we could use NumPy and a bunch of matrix operations to solve this numerically, it will be easier to use PuLP as shown in Example A-11. Note that `LpVariable` defines the variables to solve for. `LpProblem` is the linear programming system that adds constraints and objective functions using Python operators. Then the variables are solved by calling `solve()` on the `LpProblem`.

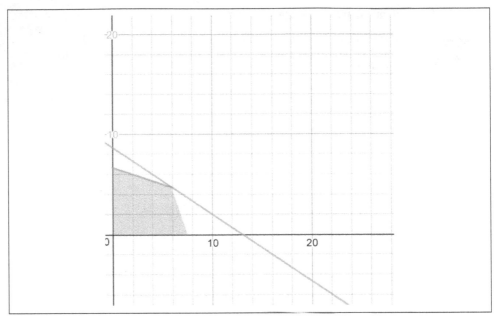

Figure A-8. Maximized objective for our linear programming system

Example A-11. Using Python PuLP to solve a linear programming system

```
# GRAPH" https://www.desmos.com/calculator/iildqi2vt7

from pulp import *

# declare your variables
x = LpVariable("x", 0)    # 0<=x
y = LpVariable("y", 0) # 0<=y

# defines the problem
prob = LpProblem("factory_problem", LpMaximize)

# defines the constraints
prob += x + 3*y <= 20
prob += 6*x +2*y <= 45

# defines the objective function to maximize
prob += 200*x + 300*y

# solve the problem
status = prob.solve()
print(LpStatus[status])

# print the results x = 5.9375, y = 4.6875
print(value(x))
print(value(y))
```

You might be wondering if it makes sense to build 5.9375 and 4.6875 units. Linear programming systems are much more efficient if you can tolerate continuous values in your variables, and perhaps you can just round them afterward. But certain types of problems absolutely require integers and binary variables to be handled discretely.

To force the x and y variables to be treated as integers, provide a category argument `cat=LpInteger` as shown in Example A-12.

Example A-12. Forcing variables to be solved as integers

```
# declare your variables
x = LpVariable("x", 0, cat=LpInteger) # 0<=x
y = LpVariable("y", 0, cat=LpInteger) # 0<=y
```

Graphically, this means we fill our feasible region with discrete points rather than a continuous region. Our solution will not land on a vertex necessarily but rather the point that is closest to the vertex as shown in Figure A-9.

There are a couple of special cases in linear programming, as shown in Figure A-10. Sometimes there can be many solutions. At times, there may be no solution at all.

This is just a quick introductory example to linear programming, and unfortunately there is not enough room in this book to do the topic justice. It can be used for surprising problems, including scheduling constrained resources (like workers, server jobs, or rooms), solving Sudokus, and optimizing financial portfolios.

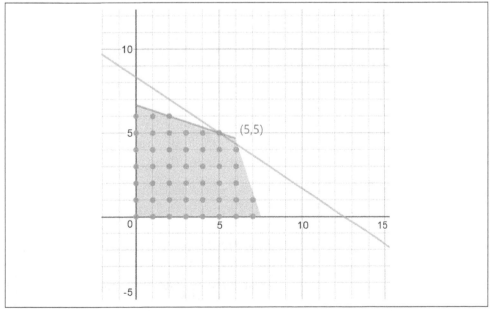

Figure A-9. A discrete linear programming system

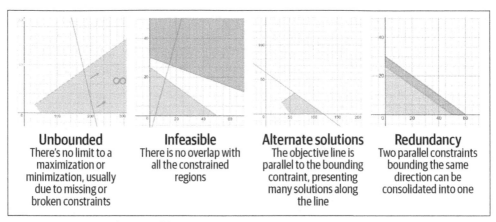

Figure A-10. Special cases of linear programming

If you want to learn more, there are some good YouTube videos out there including PatrickJMT (*https://oreil.ly/lqeeR*) and Josh Emmanuel (*https://oreil.ly/jAHWc*). If you want to deep dive into discrete optimization, Professor Pascal Van Hentenryck has done a tremendous service putting a course together on Coursera (*https://oreil.ly/aVGxY*).

MNIST Classifier Using scikit-learn

Example A-13 shows how to use scikit-learn's neural network for handwritten digit classification.

Example A-13. A handwritten digit classifier neural network in scikit-learn

```
import numpy as np
import pandas as pd
# load data
from sklearn.model_selection import train_test_split
from sklearn.neural_network import MLPClassifier

df = pd.read_csv('https://bit.ly/3ilJc2C', compression='zip', delimiter=",")

# Extract input variables (all rows, all columns but last column)
# Note we should do some linear scaling here
X = (df.values[:, :-1] / 255.0)

# Extract output column (all rows, last column)
Y = df.values[:, -1]

# Get a count of each group to ensure samples are equitably balanced
print(df.groupby(["class"]).agg({"class" : [np.size]}))

# Separate training and testing data
```

```
# Note that I use the 'stratify' parameter to ensure
# each class is proportionally represented in both sets
X_train, X_test, Y_train, Y_test = train_test_split(X, Y,
    test_size=.33, random_state=10, stratify=Y)

nn = MLPClassifier(solver='sgd',
                   hidden_layer_sizes=(100, ),
                   activation='logistic',
                   max_iter=480,
                   learning_rate_init=.1)

nn.fit(X_train, Y_train)

print("Training set score: %f" % nn.score(X_train, Y_train))
print("Test set score: %f" % nn.score(X_test, Y_test))

# Display heat map
import matplotlib.pyplot as plt
fig, axes = plt.subplots(4, 4)

# use global min / max to ensure all weights are shown on the same scale
vmin, vmax = nn.coefs_[0].min(), nn.coefs_[0].max()
for coef, ax in zip(nn.coefs_[0].T, axes.ravel()):
    ax.matshow(coef.reshape(28, 28), cmap=plt.cm.gray, vmin=.5 * vmin, vmax=.5 * vmax)
    ax.set_xticks(())
    ax.set_yticks(())

plt.show()
```

Exercise Answers

Chapter 1

1. 62.6738 is rational because it has a finite number of decimal places, and therefore can be expressed as a fraction 626738 / 10000.

2. $10^7 10^{-5} = 10^{7 + -5} = 10^2 = 100$

3. $81^{\frac{1}{2}} = \sqrt{(81)} = 9$

4. $25^{\frac{3}{2}} = \left(25^{1/2}\right)^3 = 5^3 = 125$

5. The resulting amount would be \$1,161.47. The Python script is as follows:

```
from math import exp

p = 1000
r = .05
t = 3
n = 12

a = p * (1 + (r/n))**(n * t)

print(a) # prints 1161.4722313334678
```

6. The resulting amount would be \$1161.83. The Python script is as follows:

```
from math import exp

p = 1000 # principal, starting amount
r = .05 # interest rate, by year
t = 3.0 # time, number of years

a = p * exp(r*t)
```

```
print(a) # prints 1161.834242728283
```

7. The derivative calculates to $6x$, which would make the slope at $x = 3$ to be 18. The SymPy code is as follows:

```
from sympy import *

# Declare 'x' to SymPy
x = symbols('x')

# Now just use Python syntax to declare function
f = 3*x**2 + 1

# Calculate the derivative of the function
dx_f = diff(f)
print(dx_f) # prints 6*x
print(dx_f.subs(x,3)) # 18
```

8. The area under the curve between 0 and 2 is 10. The SymPy code is as follows:

```
from sympy import *

# Declare 'x' to SymPy
x = symbols('x')

# Now just use Python syntax to declare function
f = 3*x**2 + 1

# Calculate the integral of the function with respect to x
# for the area between x = 0 and 2
area = integrate(f, (x, 0, 2))

print(area) # prints 10
```

Chapter 2

1. $0.3 \times 0.4 = 0.12$; refer to "Joint Probabilities" on page 44.

2. $(1 - 0.3) + 0.4 - ((1 - 0.3) \times 0.4) = 0.82$; refer to "Union Probabilities" on page 45 and remember we are looking for *NO RAIN* so subtract that probability from 1.0.

3. $0.3 \times 0.2 = 0.06$; refer to "Conditional Probability and Bayes' Theorem" on page 47.

4. The following Python code calculates an answer of 0.822, adding up the probabilities of 50 or more passengers who do not show up:

```
from scipy.stats import binom

n = 137
p = .40
```

```
p_50_or_more_noshows = 0.0

for x in range(50,138):
    p_50_or_more_noshows += binom.pmf(x, n, p)

print(p_50_or_more_noshows) # 0.822095588147425
```

5. Using the beta distribution shown in the following SciPy code, get the area up to 0.5 and subtract it from 1.0. Result is about 0.98, so this coin is highly unlikely to be fair.

```
from scipy.stats import beta

heads = 8
tails = 2

p = 1.0 - beta.cdf(.5, heads, tails)

print(p) # 0.98046875
```

Chapter 3

1. The mean is 1.752 and the standard deviation is approximately 0.02135. The Python code is as follows:

```
from math import sqrt

sample = [1.78, 1.75, 1.72, 1.74, 1.77]

def mean(values):
    return sum(values) /len(values)

def variance_sample(values):
    mean = sum(values) / len(values)
    var = sum((v - mean) ** 2 for v in values) / len(values)
    return var

def std_dev_sample(values):
    return sqrt(variance_sample(values))

mean = mean(sample)
std_dev = std_dev_sample(sample)

print("MEAN: ", mean) # 1.752
print("STD DEV: ", std_dev) # 0.02135415650406264
```

2. Use the CDF to get the value between 30 and 20 months, which is an area of about 0.06. The Python code is as follows:

```python
from scipy.stats import norm

mean = 42
std_dev = 8

x = norm.cdf(30, mean, std_dev) - norm.cdf(20, mean, std_dev)

print(x) # 0.06382743803380352
```

3. There's a 99% probability the average filament diameter for a roll is between 1.7026 and 1.7285. The Python code is as follows:

```python
from math import sqrt
from scipy.stats import norm

def critical_z_value(p, mean=0.0, std=1.0):
    norm_dist = norm(loc=mean, scale=std)
    left_area = (1.0 - p) / 2.0
    right_area = 1.0 - ((1.0 - p) / 2.0)
    return norm_dist.ppf(left_area), norm_dist.ppf(right_area)


def ci_large_sample(p, sample_mean, sample_std, n):
    # Sample size must be greater than 30

    lower, upper = critical_z_value(p)
    lower_ci = lower * (sample_std / sqrt(n))
    upper_ci = upper * (sample_std / sqrt(n))

    return sample_mean + lower_ci, sample_mean + upper_ci


print(ci_large_sample(p=.99, sample_mean=1.715588,
    sample_std=0.029252, n=34))
# (1.7026658973748656, 1.7285101026251342)
```

4. The marketing campaign worked with a p-value of 0.01888. The Python code is as follows:

```python
from scipy.stats import norm

mean = 10345
std_dev = 552

p1 = 1.0 - norm.cdf(11641, mean, std_dev)

# Take advantage of symmetry
p2 = p1

# P-value of both tails
# I could have also just multiplied by 2
p_value = p1 + p2
```

```
print("Two-tailed P-value", p_value)
if p_value <= .05:
    print("Passes two-tailed test")
else:
    print("Fails two-tailed test")

# Two-tailed P-value 0.01888333596496139
# Passes two-tailed test
```

Chapter 4

1. Vector lands on [2, 3]. The Python code is as follows:

```
from numpy import array

v = array([1,2])

i_hat = array([2, 0])
j_hat = array([0, 1.5])

# fix this line
basis = array([i_hat, j_hat])

# transform vector v into w
w = basis.dot(v)

print(w) # [2, 3]
```

2. Vector lands on [0, -3]. The Python code is as follows:

```
from numpy import array

v = array([1,2])

i_hat = array([-2, 1])
j_hat = array([1, -2])

# fix this line
basis = array([i_hat, j_hat])

# transform vector v into w
w = basis.dot(v)

print(w) # [ 0, -3]
```

3. Determinant is 2.0. The Python code is as follows:

```
import numpy as np
from numpy.linalg import det

i_hat = np.array([1, 0])
j_hat = np.array([2, 2])

basis = np.array([i_hat,j_hat]).transpose()

determinant = det(basis)

print(determinant) # 2.0
```

4. Yes, because matrix multiplication allows us to combine several matrices into a single matrix representing one consolidated transformation.

5. $x = 19.8$, $y = -5.4$, $z = -6$. The code is as follows:

```
from numpy import array
from numpy.linalg import inv

A = array([
    [3, 1, 0],
    [2, 4, 1],
    [3, 1, 8]
])

B = array([
    54,
    12,
    6
])

X = inv(A).dot(B)

print(X) # [19.8 -5.4 -6. ]
```

6. Yes, it is linearly dependent. Although we have some floating point imprecision with NumPy, the determinant is effectively 0:

```
from numpy.linalg import det
from numpy import array

i_hat = array([2, 6])
j_hat = array([1, 3])

basis = array([i_hat, j_hat]).transpose()
print(basis)

determinant = det(basis)

print(determinant) # -3.330669073875464e-16
```

To get around the floating point issues, use SymPy and you will get 0:

```
from sympy import *

basis = Matrix([
    [2,1],
    [6,3]
])

determinant = det(basis)

print(determinant) # 0
```

Chapter 5

1. There are many tools and approaches to perform a linear regression as we learned in Chapter 5, but here is the solution using scikit-learn. The slope is 1.75919315 and the intercept is 4.69359655.

    ```
    import pandas as pd
    import matplotlib.pyplot as plt
    from sklearn.linear_model import LinearRegression

    # Import points
    df = pd.read_csv('https://bit.ly/3C8JzrM', delimiter=",")

    # Extract input variables (all rows, all columns but last column)
    X = df.values[:, :-1]

    # Extract output column (all rows, last column)
    Y = df.values[:, -1]

    # Fit a line to the points
    fit = LinearRegression().fit(X, Y)

    # m = 1.75919315, b = 4.69359655
    m = fit.coef_.flatten()
    b = fit.intercept_.flatten()
    print("m = {0}".format(m))
    print("b = {0}".format(b))

    # show in chart
    plt.plot(X, Y, 'o') # scatterplot
    plt.plot(X, m*X+b) # line
    plt.show()
    ```

2. We get a high correlation of 0.92421 and a test value of 23.8355 with a statistically significant range of ±1.9844. This correlation is definitely useful and statistically significant. The code is as follows:

    ```
    import pandas as pd
    ```

```
# Read data into Pandas dataframe
df = pd.read_csv('https://bit.ly/3C8JzrM', delimiter=",")

# Print correlations between variables
correlations = df.corr(method='pearson')
print(correlations)

# OUTPUT:
#            x          y
# x   1.00000   0.92421
# y   0.92421   1.00000

# Test for statistical significance
from scipy.stats import t
from math import sqrt

# sample size
n = df.shape[0]
print(n)
lower_cv = t(n - 1).ppf(.025)
upper_cv = t(n - 1).ppf(.975)

# retrieve correlation coefficient
r = correlations["y"]["x"]

# Perform the test
test_value = r / sqrt((1 - r ** 2) / (n - 2))

print("TEST VALUE: {}".format(test_value))
print("CRITICAL RANGE: {}, {}".format(lower_cv, upper_cv))

if test_value < lower_cv or test_value > upper_cv:
    print("CORRELATION PROVEN, REJECT H0")
else:
    print("CORRELATION NOT PROVEN, FAILED TO REJECT H0 ")

# Calculate p-value
if test_value > 0:
    p_value = 1.0 - t(n - 1).cdf(test_value)
else:
    p_value = t(n - 1).cdf(test_value)

# Two-tailed, so multiply by 2
p_value = p_value * 2
print("P-VALUE: {}".format(p_value))

"""
TEST VALUE: 23.835515323677328
CRITICAL RANGE: -1.9844674544266925, 1.984467454426692
CORRELATION PROVEN, REJECT H0
P-VALUE: 0.0 (extremely small)
"""
```

3. At $x = 50$, the prediction interval is between 50.79 and 134.51. The code is as follows:

```python
import pandas as pd
from scipy.stats import t
from math import sqrt

# Load the data
points = list(pd.read_csv('https://bit.ly/3C8JzrM', delimiter=",") \
    .itertuples())

n = len(points)

# Linear Regression Line
m = 1.75919315
b = 4.69359655

# Calculate Prediction Interval for x = 50
x_0 = 50
x_mean = sum(p.x for p in points) / len(points)

t_value = t(n - 2).ppf(.975)

standard_error = sqrt(sum((p.y - (m * p.x + b)) ** 2 for p in points) / \
    (n - 2))

margin_of_error = t_value * standard_error * \
                sqrt(1 + (1 / n) + (n * (x_0 - x_mean) ** 2) / \
                    (n * sum(p.x ** 2 for p in points) - \
        sum(p.x for p in points) ** 2))

predicted_y = m*x_0 + b

# Calculate prediction interval
print(predicted_y - margin_of_error, predicted_y + margin_of_error)
# 50.792086501055955 134.51442159894404
```

4. The testing datasets do moderately well when split into thirds and evaluated with k-fold, where $k = 3$. You will get a mean of roughly 0.83 in MSE and a standard deviation of 0.03 across the three datasets.

```python
import pandas as pd
from sklearn.linear_model import LinearRegression
from sklearn.model_selection import KFold, cross_val_score

df = pd.read_csv('https://bit.ly/3C8JzrM', delimiter=",")

# Extract input variables (all rows, all columns but last column)
X = df.values[:, :-1]

# Extract output column (all rows, last column)\
Y = df.values[:, -1]
```

```
# Perform a simple linear regression
kfold = KFold(n_splits=3, random_state=7, shuffle=True)
model = LinearRegression()
results = cross_val_score(model, X, Y, cv=kfold)
print(results)
print("MSE: mean=%.3f (stdev-%.3f)" % (results.mean(), results.std()))
"""
[0.86119665 0.78237719 0.85733887]
MSE: mean=0.834 (stdev-0.036)
"""
```

Chapter 6

1. The accuracy is extremely high when you run this through scikit-learn. When I run it, I get at least 99.9% accuracy on average with the test folds.

```
import pandas as pd
from sklearn.linear_model import LogisticRegression
from sklearn.metrics import confusion_matrix
from sklearn.model_selection import KFold, cross_val_score


# Load the data
df = pd.read_csv("https://bit.ly/3imidqa", delimiter=",")

X = df.values[:, :-1]
Y = df.values[:, -1]

kfold = KFold(n_splits=3, shuffle=True)
model = LogisticRegression(penalty='none')
results = cross_val_score(model, X, Y, cv=kfold)

print("Accuracy Mean: %.3f (stdev=%.3f)" % (results.mean(),
results.std()))
```

2. The confusion matrix will yield an extremely high number of true positives and true negatives, and very few false positives and false negatives. Run this code and you will see:

```
import pandas as pd
from sklearn.linear_model import LogisticRegression
from sklearn.metrics import confusion_matrix
from sklearn.model_selection import train_test_split

# Load the data
df = pd.read_csv("https://bit.ly/3imidqa", delimiter=",")

# Extract input variables (all rows, all columns but last column)
X = df.values[:, :-1]

# Extract output column (all rows, last column)\
Y = df.values[:, -1]
```

```
model = LogisticRegression(solver='liblinear')

X_train, X_test, Y_train, Y_test = train_test_split(X, Y, test_size=.33)
model.fit(X_train, Y_train)
prediction = model.predict(X_test)


"""
The confusion matrix evaluates accuracy within each category.
[[truepositives falsenegatives]
 [falsepositives truenegatives]]

The diagonal represents correct predictions,
so we want those to be higher
"""
matrix = confusion_matrix(y_true=Y_test, y_pred=prediction)
print(matrix)
```

3. An interactive shell for testing user-input colors is shown next. Consider testing black (0,0,0) and white (255,255,255) to see if dark and light fonts respectively are predicted correctly.

```
import pandas as pd
from sklearn.linear_model import LogisticRegression
import numpy as np
from sklearn.model_selection import train_test_split

# Load the data
df = pd.read_csv("https://bit.ly/3imidqa", delimiter=",")

# Extract input variables (all rows, all columns but last column)
X = df.values[:, :-1]

# Extract output column (all rows, last column)
Y = df.values[:, -1]

model = LogisticRegression(solver='liblinear')

X_train, X_test, Y_train, Y_test = train_test_split(X, Y, test_size=.33)
model.fit(X_train, Y_train)
prediction = model.predict(X_test)

# Test a prediction
while True:
    n = input("Input a color {red},{green},{blue}: ")
    (r, g, b) = n.split(",")
    x = model.predict(np.array([[int(r), int(g), int(b)]]))
    if model.predict(np.array([[int(r), int(g), int(b)]]))[0] == 0.0:
        print("LIGHT")
    else:
        print("DARK")
```

4. Yes, the logistic regression is very effective at predicting light or dark fonts for a given background color. Not only is the accuracy extremely high, but the confusion matrix has high numbers in the top-right to bottom-left diagonal with lower numbers in the other cells.

Chapter 7

There's obviously a lot of experimenting and alchemy you can try with different hidden layers, activation functions, different testing dataset sizes, and so on. I tried to use one hidden layer with three nodes with a ReLU activation, and I struggled to get good predictions on my testing dataset. The confusion matrices and accuracy were consistently poor and any configuration changes I ran did just as poorly.

The reasons the neural network is probably failing are 1) the testing dataset is far too small for a neural network (which are extremely data-hungry) and 2) there are simpler and more effective models like logistic regression for this type of problem. That's not to say you can't find a configuration that will work, but you have to be careful not to p-hack your way to a good result that overfits to the little training and testing data you have.

Here is the scikit-learn code I used:

```python
import pandas as pd
# load data
from sklearn.metrics import confusion_matrix
from sklearn.model_selection import train_test_split
from sklearn.neural_network import MLPClassifier

df = pd.read_csv('https://tinyurl.com/y6r7qjrp', delimiter=",")

# Extract input variables (all rows, all columns but last column)
X = df.values[:, :-1]

# Extract output column (all rows, last column)
Y = df.values[:, -1]

# Separate training and testing data
X_train, X_test, Y_train, Y_test = train_test_split(X, Y, test_size=1/3)

nn = MLPClassifier(solver='sgd',
                   hidden_layer_sizes=(3, ),
                   activation='relu',
                   max_iter=100_000,
                   learning_rate_init=.05)

nn.fit(X_train, Y_train)

print("Training set score: %f" % nn.score(X_train, Y_train))
print("Test set score: %f" % nn.score(X_test, Y_test))
```

```
print("Confusion matrix:")
matrix = confusion_matrix(y_true=Y_test, y_pred=nn.predict(X_test))
print(matrix)
```

Index

C

calculus (see math and calculus review)

career advice for data scientists (see data science career)

Cartesian plane, 8

CAS (computer algebra system), 24

causation versus correlation, 180

CDF (see cumulative density function)

central limit theorem, 89-92, 94, 95

chain rule, 31-33, 245-253

chi-square (chi2) distribution, 216

chi2 module (SciPy), 217

Chollet, Francis, 256

class imbalance, 225, 244

classification, 193

    class imbalance, 225, 244

    in logistic regression, 195, 219-223

    MNIST classifier, 308

    and neural networks, 230, 233, 253-254

    pitfalls of, 207

closed form equation, 157-158

coefficient of determination (r^2), linear regression, 179, 187

coefficient of determination (R^2), logistic regression, 211-216

coefficient of variation, 89

commutative property of addition, 116

complex numbers, 3

computational complexity, 158

computer algebra system (CAS), 24

conditional probability, 47-49

confidence intervals, 92-95, 184, 185

confirmation bias, 68

confounding variables, 67

confusion matrices, 219-223

continuous versus discrete variables, 193

controlled variable, 96

convolutional neural networks, 230

coordinate plane, 8

corr() function, 172

correlation coefficient (r), 171-179

correlation matrix, 179

correlation versus causation, 180

critical t-value, 105, 176

critical z-value, 92

cross-validation of dataset, 188

The Cult of We (Brown and Farrell), 276

cumulative density function (CDF), 33

    in beta distribution, 55-58

creating from scratch, 295-296

inverse, 85-89, 92, 296

in normal distribution integration, 82-85, 295-296

curvilinear functions, 9

D

data, 64-65

    biases (see bias)

    big data considerations, 105-106, 185, 263

    operations in vector terms, 118

    politics of access to, 280

    questioning sources and assumptions, 65, 207

data mining, 64, 105, 106

data privacy issue, 207

data science

    historical sketch, 262-264

    linear algebra as foundation of, 145

    linear transformations' role in, 131

    machine learning (see machine learning)

    redefining, 261-264

    state of the field, ix-x

    statistics (see statistics)

data science career

    development of field, 262-264

    disciplines included in, 261

    hard and soft skills needed, 265-276

    importance of boundary setting in, 286

    job market considerations, 277-285

    labor market, managing in current, 286

    tools included in, 261

Data Science from Scratch (Grus), 270

data silo breakdown, as IT job, not data science job, 279

deep learning, 229

    in development of 21st-century data science, 264

    limitations of, 255-258

    when to use, 230

Deep Learning with Python (Chollet), 256

degrees of freedom (df) parameter, 105, 176, 181, 216

dependent variables, 6

derivatives, 24-33

    chain rule, 31-33

    in gradient descent for linear regression, 163-166

    partial, 28-29

inverse cumulative density function (CDF), 85-89, 92, 296
inverse matrices, 136, 138-142, 158-161
inversion, linear transformation, 124
irrational numbers, 3, 16
IT department
    data scientists being siphoned off to, 281, 284
    shadow IT issue, 284

J

Jabberwocky Effect, 277, 282
Java data science libraries, 270
JavaScript D3.js, 271
JAX library, 250
job market for data scientists, 277-285
    adequate resources, 280
    competing with existing systems, 282-284
    organizational focus and buy-in, 278-280
    realistic expectations, 285
    reasonable objectives, 281-282
    role definition, 277
    shadow IT, 284
    unexpected role challenge, 284
joint conditional probability, 49
joint probability, 44-45, 200
Jupyter Notebooks, 269

K

k-fold validation, train/test splits, 189
Kozyrkov, Cassie, 263, 283

L

lambdify() function, 166
lasso regression, 169
LaTeX rendering with SymPy, 289-290
leaky ReLU function, 236
learning rate, machine learning and gradient descent, 161
Learning SQL, 3rd Edition (Beaulieu), 266
least sum of squares, calculating, 157
leave-one-out cross-validation (LOOCV), 189
level of confidence (LOC), 92
likelihood, compared to probability, 42
limits, mathematical/calculus, 22-24, 29-31, 38
linear algebra, 109-146
    determinants, 120, 131-135
    eigenvectors and eigenvalues, 142-145

as foundation of data science, 145
inverse matrices, 138-142, 158-161
linear transformations, 121-129
matrix multiplication, 129-131
special matrix types, 136-138
systems of equations, 138-142
vectors, 110-120
linear correlation, 149
linear dependence, vectors, 119-120, 134
linear independence, 120
linear programming, 142, 145
linear regression, 147-192, 302-307
    basic, 149-153
    benefits and costs, 150
    closed form equation, 157-158
    coefficient of determination, 179
    correlation coefficient, 171-179
    fitting regression line to data, 153-166
    gradient descent, 161-166
    hill climbing algorithm, 298-300
    image classification neural network, 233
    inverse matrix techniques, 158-161
    multiple linear regression, 191
    overfitting and variance, 167-169
    prediction intervals, 181-184
    residuals and squared errors, 153-157
    standard error of the estimate, 180
    statistical significance, 174-179
    stochastic gradient descent, 169-170
    train/test splits, 185-190
linear transformations, 121-129
    basis vectors, 121-124
    matrix vector multiplication, 124-129
LinearRegression() function, 198
ln(), natural logarithm, 21
LOC (level of confidence), 92
log likelihood of fit, calculating, 212-214
log() function, 16, 202
log-odds (logit) function, 196, 208-211
logarithmic addition, 202
logarithms, 16-17 (see logistic regression)
logistic activation function, 235, 236, 239
logistic function, 195-197
logistic regression, 193-227
    class imbalance, 225, 244
    confusion matrices, 219-223
    fitting logistic curve, 198-204
    hill climbing algorithm, 301-302
    log-odds, 208-211

convolutional, 230
forward propagation, 239-244
limitations of, 255-258
MNIST classifier, 238, 308
recurrent, 230
regression's role in, 233, 264
scikit-learn, 253-254, 308
when to use, 230
Ng, Andrew, 190, 255
Nield, Thomas, 266
No Bullshit Guide to Math and Physics (Savov),
    1
nonmutually exclusive events, probabilities,
    45-47
norm.ppf() function, 85
normal distribution, 78-85
    and central limit theorem, 89-92, 94, 95
    and chi-square (chi2) distribution, 216
    integrating with CDF, 82-85, 295-296
    linear regression and prediction intervals,
        181-182
    probability density function, 81
    properties, 81
    random number generation from, 86
    versus T-distribution, 104
    Z-scores, 87
NoSQL, 266
null hypothesis (H0), 96, 97-103
number theory, 2-3
numerical stability, 161
NumPy
    decomposition for linear regression, 160
    determinant calculation, 132
    eigenvector and eigenvalue calculation, 143
    inverse and transposed matrices, 159
    matrix decomposition and recomposition,
        144
    matrix multiplication, 129
    matrix vector multiplication, 125-127
    scaling a number, 117
    stochastic gradient descent, 169
    vector addition, 114
    vector declaration, 111, 113
    vector transformation, 125

O

odds
    probability expressed as, 42
    relationship to logarithm, 209

one-tailed test, 97-99
order of operations, 3-5
outlier values, and use of median, 72
output variables, 6
overfitting of data
    linear regression, 167-169
    machine learning technique to mitigate,
        185-190
    neural networks' challenge of, 255-257
    train/test split to mitigate, 185-190
    versus underfitting, 169

P

p-hacking, 103
p-values, 95
    linear regression significance test, 178
    logistic regression, 216-218
    for one-tailed test, 99
    scikit-learn's lack of, 185
    for two-tailed test, 101, 103
Pandas, 152
    closed form equation, 157-158
    correlation coefficient calculation, 172
    creating correlation matrix, 179
    gradient descent for linear regression, 163
    importing SQL query into, 266
    log likelihood of fit, 212-214
    loss function for linear regression, 166
    residuals calculation, 154
    standard error of estimate calculation, 181
    stochastic gradient descent, 169
    sum of squares calculation, 155
parabola, 9
partial derivatives, 28-29
PatrickJMT, 142, 308
PCA (principal component analysis), 143
PDF (probability density function), 81, 83
Pearson correlation (see correlation coefficient)
Plotly, 152, 271
populations, 66-67
    and central limit theorem, 89-92
    mean calculation, 70, 92
    and standard deviation, 75
    and variance, 73-76
positive correlation, 172
power rule, exponents, 15
ppf() function, 85
practitioner versus advisor roles, deciding on,
    274-276

predict() function, 199
prediction
    statistical significance, 1
prediction intervals, 181-184
predict_prob() function, 199
predict_proba() function, 206
principal component analysis (PCA), 143
privacy, data, 207
probability, 41-60
    beta distribution, 53-60
    binomial distribution, 51-53
    conditional, 47-49
    joint, 44-45, 200
    joint conditional, 49
    versus likelihood, 42
    logistic regression (see logistic regression)
    marginal, 44
    and p-value, 95
    permutations and combinations, 44
    and statistics, 43, 63
    sum rule of, 46
    union, 45-47
    union conditional, 50
probability density function (PDF), 81, 83
problem versus tool, starting with, 257, 259, 276
product rule
    exponents, 13
    probabilities, 45
profitability and defining success, 276
PuLP, 112, 302, 305
Pyomo, 302
Python, 1
    (see also specific tools)
    central limit theorem, 90
    continuous interest calculation, 20-21
    converting x-values into Z-scores, 88
    correlation coefficient from scratch, 173
    derivative calculator, 26
    forward propagation network, 239
    hill-climbing algorithms, 299, 301-302
    integral approximation, 36-37
    interactive shell, adding to neural network, 252
    joint likelihood calculation, 201
    libraries used, x
    log() function, 17
    logarithmic addition, 202
    logistic function, 196

mean calculation, 70
median calculation, 71
mode calculation, 73
natural logarithm, 21
normal distribution function, 81, 295
plotting function graphs, 8
solving numerical expression, 4
speed compared to numerical libraries, 111
SQL query from, 266
standard deviation calculation, 75, 77
stochastic gradient descent, 250-252
summation of elements, 12
variables in, 5
variance calculation, 75
weighted mean calculation, 71
Python for Data (McKinney), 146

Q

QR decomposition method, 160
quantile, 72
quartiles, 72

R

R-squared (R^2), 211-216
Ramalho, Luciano, 268
random numbers, generating from normal distribution, 86
random selection of data sample, 67
random-fold validation, 189
randomness, 171
    accounting for, 105-106, 178
    null hypothesis (H0), 96, 97-103
range() function, 12
rational numbers, 2
real numbers, 3, 7
real-world versus simulation data, neural network challenges, 256
receiver operator characteristic (ROC) curve, 223-225
recurrent neural networks, 230
redundancy case, linear programming, 307
regression, 147
    (see also linear regression; logistic regression)
Reimann Sums, 38
relational database, 265
ReLU activation function, 234, 236, 239
residuals, 153-157
RGB color values, 231-233

ridge regression, 169
Riemann Sums, 33
ROC (receiver operator characteristic) curve, 223-225
role specialization in data science career, 286
rotation, linear transformation, 124

S

samples, statistical, 66-67
    bias, 67, 69, 77, 151
    and central limit theorem, 89-92, 95
    mean calculation, 70, 92
    random selection of, 67
    size of, 92
    standard deviation, 76-78
    T-distribution for small samples, 95, 104-105, 176-179
    variance calculation, 76-78
Savov, Ivan, 1
Scala, 271
scalar value, 116-117
scaling
    determinants in, 132-135
    linear transformation, 124
    of vectors, 116-119
scikit-learn
    AUC as parameter in, 225
    cross validation for linear regression, 188
    lack of confidence intervals and p-values in, 185
    linear regressions with, 152
    MNIST classifier, 308
    multivariable linear regression, 191
    multivariable logistic regression, 205
    neural network classifier, 253-254
    random-fold validation for linear regression, 189
    three-fold cross-validation logistic regression, 219
    train/test split on linear regression, 186
SciPy
    basic linear regression, 151-153
    beta distribution, 56-60
    binomial distribution calculation, 52-53
    confidence interval calculation, 94
    confusion matrix for test dataset, 221
    critical value from T-distribution, 176
    critical z-value retrieval, 93
    fitting regression line to data, 152

inverse CDF, 85
    logistic regression documentation, 200
    maximum likelihood estimation, 198-200
    normal distribution CDF, 84
    p-value calculations, 99, 102, 217
    prediction interval calculation, 183
    T-distribution calculation, 105
    testing significance for linear-looking data, 177
    for x-value with 5% behind it, 98
Seaborn, for data visualization, 271
self-selection bias, 67, 68
shadow IT, 284
sigmoid curve, 196
skill sets for data science career, 265-276
    data visualization, 271
    knowing your industry, 272
    practitioner versus advisor, 274-276
    productive learning, 274
    programming proficiency, 268-271
    SQL proficiency, 265-267
small datasets
    T-distribution, 95, 104-105, 176-179
    and train/test splits, 186, 189
Smith, Gary, 105
SMOTE algorithms, 226
Softmax function, 236
software engineering, role in data science, 264, 268
software licenses, 272
span and linear dependence, vectors, 119-120
sparse matrix, 138
SQL (structured query language), 265-267
SQL Pocket Guide, 4th Edition (Zhao), 266
sqrt() function, 75
square matrix, 136, 143
squared errors, 153-157
standard deviation
    normal distribution role of, 82
    for population, 75
    sample calculation of, 76-78
    and standard error of the estimate, 181
Standard Deviations (Smith), 105
standard error of the estimate (S_e), 181
standard normal distribution, 87
Starmer, Josh, 77, 209, 226
statistical learning, 148
statistical significance
    inferential statistics, 96-103

About the Author

Thomas Nield is the founder of Nield Consulting Group as well as an instructor at O'Reilly Media and the University of Southern California. He enjoys making technical content relatable and relevant to those unfamiliar with or intimidated by it. Thomas regularly teaches classes on data analysis, machine learning, mathematical optimization, AI system safety, and practical artificial intelligence. He's authored two books, *Getting Started with SQL* (O'Reilly) and *Learning RxJava* (Packt). He's also the founder and inventor of Yawman Flight, a company that develops universal handheld controls for flight simulation and unmanned aerial vehicles.

Colophon

The animals on the cover of *Essential Math for Data Science* are four-striped grass mice (*Rhabdomys pumilio*). These rodents are found in the southern half of the African continent, in varied habitats such as savanna, desert, farmland, shrublands, and even cities. As its common name suggests, this animal has a distinct set of four dark stripes running down its back. Even at birth, these stripes are visible as pigmented lines in the pup's hairless skin.

The coloring of the grass mouse's fur varies from dark brown to grayish white, with lighter sides and bellies. In general, the species grows to about 18–21 centimeters long (not counting the tail, which is roughly equal to body length) and weighs 30–55 grams. The mouse is most active during the day, and has an omnivorous diet of seeds, plants, and insects. In the summer months, it tends to eat more plant and seed material, and it maintains fat stores to see itself through times of limited food supply.

Four-striped grass mice are easy to observe given their wide range, and have been noted to switch between solitary and social lifestyles. During the breeding season, they tend to stay separate (perhaps to avoid excessive reproductive competition) and females are territorial of their burrows. Outside of that, however, the mice congregate in groups to forage, avoid predators, and huddle together for warmth.

Many of the animals on O'Reilly covers are endangered; all of them are important to the world.

The cover illustration is by Karen Montgomery, based on an antique engraving from *The Museum of Natural History*. The cover fonts are Gilroy Semibold and Guardian Sans. The text font is Adobe Minion Pro; the heading font is Adobe Myriad Condensed; and the code font is Dalton Maag's Ubuntu Mono.

O'REILLY®

Learn from experts.
Become one yourself.

Books | Live online courses
Instant answers | Virtual events
Videos | Interactive learning

Get started at oreilly.com.

Printed in the USA
CPSIA information can be obtained
at www.ICGtesting.com
JSHW060050061024
71100JS00005BA/38